Developmental Mathematics:

A Modular Curriculum for North Carolina
Second Printing

OPERATIONS WITH INTEGERS
DMA 010

ALAN S. TUSSY
CITRUS COLLEGE

R. DAVID GUSTAFSON
ROCK VALLEY COLLEGE

DIANE R. KOENIG
ROCK VALLEY COLLEGE

BROOKS/COLE
CENGAGE Learning

Brazil • Japan • Korea • Mexico • Singapore • Spain • United Kingdom • United States

BROOKS/COLE
CENGAGE Learning

Developmental Mathematics: A Modular Curriculum for North Carolina, Second Printing: Operations with Integers
Alan S. Tussy, R. David Gustafson, Diane R. Koenig

Publisher: Charlie Van Wagner

Senior Developmental Editor: Danielle Derbenti

Senior Development Editor for Market Strategies: Rita Lombard

Assistant Editor: Stefanie Beeck

Editorial Assistant: Jennifer Cordoba

Media Editor: Heleny Wong

Marketing Manager: Gordon Lee

Marketing Assistant: Angela Kim

Marketing Communications Manager: Katy Malatesta

Content Project Manager: Jennifer Risden

Creative Director: Rob Hugel

Art Director: Vernon Boes

Print Buyer: Linda Hsu

Rights Acquisitions Account Manager, Text: Mardell Glinksi-Schultz

Rights Acquisitions Account Manager, Image: Don Schlotman

Text Designer: Diane Beasley

Photo Researcher: Bill Smith Group

Illustrators: Lori Heckelman; Graphic World Inc; Integra Software Services

Cover Designers: Ryan and Susan Stranz

Cover Image: Background: © Hemera/Thinkstock. © iStockphoto/Thinkstock.

Compositor: Integra Software Services

ISBN-13: 978-1-285-13171-9

ISBN-10: 1-285-13171-1

Brooks/Cole
20 Davis Drive
Belmont, CA 94002-3098
USA

Cengage Learning is a leading provider of customized learning solutions with office locations around the globe, including Singapore, the United Kingdom, Australia, Mexico, Brazil, and Japan. Locate your local office at **www.cengage.com/global**

Cengage Learning products are represented in Canada by Nelson Education, Ltd.

To learn more about Brooks/Cole, visit **www.cengage.com/brookscole**

Purchase any of our products at your local college store or at our preferred online store **www.CengageBrain.com**

Printed in the United States of America
1 2 3 4 5 6 7 14 13

To my lovely wife, Liz,

thank you for your insight and encouragement

ALAN S. TUSSY

To my grandchildren:

Daniel, Tyler, Spencer, Skyler, Garrett, and Jake Gustafson

R. DAVID GUSTAFSON

To my husband and my best friend, Brian Koenig

DIANE R. KOENIG

PREFACE

Developmental Mathematics: A Modular Curriculum for North Carolina is a fully integrated learning system that has been aligned to the redesigned curriculum established by the North Carolina Developmental Math Redesign Task Force. With the helpful input from instructors across the state, we have put together a program that presents problems in a meaningful context and explains the "why" behind problem solving in order to promote conceptual and sound mathematical learning. This is one of eight modules for the DMA curriculum, and is supported by a highly customizable online homework system that includes assessment tools, personalized study plans, and algorithmically generated problems to reinforce learning.

One central goal of the North Carolina Redesign Task Force was to create a curriculum with streamlined content in a modular format that could be completed in one academic year. Students can purchase only the modules needed for their developmental math requirements, and can work at a pace that is appropriate for their needs. Instructors can easily use this content with different classroom delivery methods, including self-paced Emporium labs, seated courses, and online or hybrid settings.

Another principle of the new curriculum is to develop students' conceptual understanding of mathematics through the use of contextually based problems. To that end, we have added the following features:

- New **Applied Introductions** have been written to introduce sections that are more applications-driven.
- Within the *Study Sets*, **Applications** problems and examples have been added and written to align with the NCCCS learning outcomes.
- **Concept Extensions** have been written and added to the *Study Sets* to ensure that key concepts meet the NCCCS curriculum.

In addition to new conceptual features that we have written specifically for North Carolina, we have added the following features to help guide students toward mastery of each module:

- **Course Information Sheets** start each module. These offer an explanation of the NCCCS process and ask questions that guide students to the practical knowledge that they will need in order to complete the program.
- **Are You Ready?** quizzes have been added to the beginning of each section to test students on the basic skills they will need in order to be successful with that section.
- **Module Tests**, appearing at the end of each module, have been carefully constructed to include the NCCCS learning outcomes required to pass the mastery test.

All content in these modules is supported by a corresponding prebuilt course in Enhanced WebAssign®, Cengage Learning's powerful online homework solution. Enhanced WebAssign® (EWA) engages students with immediate feedback on algorithmically generated versions of problems for unlimited practice. The *Show My Work* feature allows students to upload a file with the problem worked out, or to use a simple math palette to show their steps–helping you assess whether they understand the steps to solving a problem. The North Carolina EWA course has been prebuilt with a Personalized Study Plan, assignments, homework, and a Pre and Post Test for every module. Instructors can use the prebuilt course as is, or can customize or add material with ease.

A corresponding and fully interactive eBook, the Cengage YouBook, is integrated into the Enhanced WebAssign® course, and offers students convenient access to all module content. This powerful eBook lets you tailor the content to fit your course and provide your students with the ultimate learning experience with note-taking, highlighting, book-marking and search capabilities. Link students to your lecture notes, audio summaries, and engage them through conceptual tutorial videos as well as YouTube clips.

Cengage Learning is committed to providing unparallel service and training for faculty.

- **TeamUP Faculty Programs** help you reach and engage students by offering peer-to-peer consulting on curriculum and assessment, workshops, and professional development conferences.

TeamUP Faculty Program Consultants are a team of educators who understand your challenges whether your classroom is on-ground, online, or both.

Cengage Learning's team of **Faculty Advisors** are full-time educators and expert teachers in a diverse range of subject areas. They are available to share their experience on using Cengage Learning solutions and instructional best practices developed in their own classroom.

Explore all the ways TeamUP Faculty Programs can help you launch a new program or support your continuous improvement efforts. http://www.cengage.com/teamup/programs/ offers service and support from a dedicated team of experts to ensure your success using Enhanced WebAssign, including help with course set up, and more. http://www.cengage.com/coursecare/

TRUSTED FEATURES

- **Study Sets** found in each section offer a multifaceted approach to practicing and reinforcing the concepts taught in each section. They are designed for students to methodically build their knowledge of the section concepts, from basic recall to increasingly complex problem solving, through reading, writing, and thinking mathematically.

  Vocabulary—Each *Study Set* begins with the important *Vocabulary* discussed in that section. The fill-in-the-blank vocabulary problems emphasize the main concepts taught in the chapter and provide the foundation for learning and communicating the language of algebra.

  Concepts—In *Concepts,* students are asked about the specific subskills and procedures necessary to successfully complete the *Guided Practice* and *Try It Yourself* problems that follow.

  Notation—In *Notation,* the students review the new symbols introduced in a section. Often, they are asked to fill in steps of a sample solution. This strengthens their ability to read and write mathematics and prepares them for the *Guided Practice* problems by modeling solution formats.

  Guided Practice—The problems in *Guided Practice* are linked to an associated worked example or objective from that section. This feature promotes student success by referring them to the proper examples if they encounter difficulties solving homework problems.

  Try It Yourself—To promote problem recognition, the *Try It Yourself* problems are thoroughly mixed and are *not* linked to worked examples, giving students an opportunity to practice decision-making and strategy selection as they would when taking a test or quiz.

Applications—The *Applications* provide students the opportunity to apply their newly acquired algebraic skills to relevant and interesting real-life situations.

Writing—The *Writing* problems help students build mathematical communication skills.

Review—The *Review* problems consist of randomly selected problems from previous chapters. These problems are designed to keep students' successfully mastered skills up-to-date before they move on to the next section.

- **Detailed Author Notes** that guide students along in a step-by-step process appear in the solutions to every worked example.
- **Think It Through** features make the connection between mathematics and student life. These relevant topics often require algebra skills from the chapter to be applied to a real-life situation. Topics include tuition costs, student enrollment, job opportunities, credit cards, and many more.
- **Using Your Calculator** is an optional feature that is designed for instructors who wish to use calculators as part of the instruction in this course. This feature introduces keystrokes and shows how scientific and graphing calculators can be used to solve problems. In the *Study Sets,* icons are used to denote problems that may be solved using a calculator.

ACKNOWLEDGMENTS

We want to express our gratitude to all those who helped with this project: Steve Odrich, Mary Lou Wogan, Paul McCombs, Maria H. Andersen, Sheila Pisa, Laurie McManus, Alexander Lee, Ed Kavanaugh, Karl Hunsicker, Cathy Gong, Dave Ryba, Terry Damron, Marion Hammond, Lin Humphrey, Doug Keebaugh, Robin Carter, Tanja Rinkel, Bob Billups, Jeff Cleveland, Jo Morrison, Sheila White, Jim McClain, Paul Swatzel, Matt Stevenson, Carole Carney, Joyce Low, Rob Everest, David Casey, Heddy Paek, Ralph Tippins, Mo Trad, Eagle Zhuang, and the Citrus College library staff (including Barbara Rugeley) for their help with this project. Your encouragement, suggestions, and insight have been invaluable to us.

We would also like to express our thanks to the Cengage Learning editorial, marketing, production, and design staff for helping us craft this new edition: Danielle Derbenti, Michael Stranz, Kim Fry, Heleny Wong, Charlie Van Wagner, Jill Staut, Liz Kendall, Marc Bove, Gordon Lee, Rita Lombard, Angela Hodge, Angela Kim, Maureen Ross, Jennifer Risden, Vernon Boes, Diane Beasley, Carol O'Connell, Graphic World and Integra Software Services.

Additionally, we would like to say that authoring a textbook is a tremendous undertaking. Producing a product of this scale that is customized to match a brand new curriculum would not have been possible without the thoughtful feedback and support from the following colleagues from throughout North Carolina listed below. Their contributions to this edition have shaped the creation of this book in countless ways.

A special acknowledgment is due to Lisa Key Brown, of Central Carolina Community College. Lisa's experience in the Developmental Math classroom, detailed knowledge of the new North Carolina curriculum, and expertise in using Enhanced WebAssign has been invaluable to us as we have prepared this developmental math program.

Alan S. Tussy
R. David Gustafson
Diane R. Koenig

North Carolina Developmental Mathematics Advisory Board

Kochi Angar, *Nash Community College*
Tim Beaver, *Isothermal Community College*
Lisa Key Brown, *Central Carolina Community College*
Ann DeBoever, *Catawba Valley Community College*
Robin Hendrix, *Surry Community College*
Brenda C. Keller, *Central Carolina Community College*
Susan Lauer, *South Piedmont Community College*
Judy McInerney, *Sandhills Community College*
Nancy Paxton, *Sandhills Community College*
Molly Jane Ray, *Vance-Granville Community College*
Karen H. Sanders, *Piedmont Community College*
Calvin Stansbury, *Halifax Community College*
Elaine Tucker, *Sandhills Community College*
Solomon L. Willis, *Cleveland Community College*
Susan Worth, *Surry Community College*
Sue Zimmerman, *Vance-Granville Community College*

Content Reviewers

Darla Aguilar, *Pima Community College*
Sheila Anderson, *Housatonic Community College*
David Behrman, *Somerset Community College*
Michael Branstetter, *Hartnell College*
Joseph A. Bruno, Jr., *Community College of Allegheny County*
Eric Compton, *Brookdale Community College*
Joy Conner, *Tidewater Community College*
Ruth Dalrymple, *Saint Philip's College*
Nathalie Darden, *Brookdale Community College*
John D. Driscoll, *Middlesex Community College*
LaTonya Ellis, *Bishop State Community College*
Steven Felzer, *Lenoir Community College*
Rhoderick Fleming, *Wake Technical Community College*
Heather Gallacher, *Cleveland State University*
Kathirave Giritharan, *John A. Logan College*
Marilyn Green, *Merritt College and Diablo Valley College*
Joseph W. Giuciardi, *Community College of Allegheny County*
Geoff Hagopian, *College of the Desert*
Deborah Hanus, *Brookhaven College*
A.T. Hayashi, *Oxnard College*
Cheryl Hobneck, *Illinois Valley Community College*
Todd J. Hoff, *Wisconsin Indianhead Technical College*
Susan Kautz, *Cy-Fair College*
Jack Keating, *Massasoit Community College*
Russ Alan Killingsworth, *Seattle Pacific University*
Sandy Lofstock, *Saint Petersburg College–Tarpon Springs*
Lynn Marecek, *Santa Ana College*
Lois Martin, *Massasoit Community College*
Mikal McDowell, *Cedar Valley College*
Chris Mirbaha, *The Community College of Baltimore County*
K. Maggie Pasqua, *Brookdale Community College*
Gregory Perkins, *Hartnell College*
Euguenia Peterson, *City Colleges of Chicago–Richard Daley*
Carol Ann Poore, *Hinds Community College*
Christopher Quarles, *Shoreline Community College*
George Reed, *Angelina College*

Patricia C. Rome, *Delgado Community College*
Patricia B. Roux, *Delgado Community College*
Rebecca Rozario, *Brookdale Community College*
John Squires, *Cleveland State Community College*
Sharon Testone, *Onondaga Community College*
Bill Thompson, *Red Rocks Community College*
Barbara Tozzi, *Brookdale Community College*
Donna Tupper, *Community College of Baltimore County–Essex*
Andreana Walker, *Calhoun Community College*
Jane Wampler, *Housatonic Community College*
Arminda Wey, *Brookdale Community College*
Mary Lou Wogan, *Klamath Community College*
Valerie Wright, *Central Piedmont Community College*
Kevin Yokoyama, *College of the Redwoods*
Mary Young, *Brookdale Community College*

ABOUT THE AUTHORS

Alan S. Tussy

Alan Tussy teaches all levels of developmental mathematics at Citrus College in Glendora, California. He has written nine math books—a paperback series and a hardcover series. A meticulous, creative, and visionary teacher who maintains a keen focus on his students' greatest challenges, Alan Tussy is an extraordinary author, dedicated to his students' success. Alan received his Bachelor of Science degree in Mathematics from the University of Redlands and his Master of Science degree in Applied Mathematics from California State University, Los Angeles. He has taught up and down the curriculum from Prealgebra to Differential Equations. He is currently focusing on the developmental math courses. Professor Tussy is a member of the American Mathematical Association of Two-Year Colleges.

R. David Gustafson

R. David Gustafson is Professor Emeritus of Mathematics at Rock Valley College in Illinois and coauthor of several best-selling math texts, including Gustafson/Frisk's *Beginning Algebra, Intermediate Algebra, Beginning and Intermediate Algebra: A Combined Approach, College Algebra,* and the Tussy/Gustafson developmental mathematics series. His numerous professional honors include Rock Valley Teacher of the Year and Rockford's Outstanding Educator of the Year. He earned a Master of Arts from Rockford College in Illinois, as well as a Master of Science from Northern Illinois University.

Diane R. Koenig

Diane Koenig received a Bachelor of Science degree in Secondary Math Education from Illinois State University in 1980. She began her career at Rock Valley College in 1981, when she became the Math Supervisor for the newly formed Personalized Learning Center. Earning her Master's Degree in Applied Mathematics from Northern Illinois University, Ms. Koenig in 1984 had the distinction of becoming the first full-time woman mathematics faculty member at Rock Valley College. In addition to being nominated for AMATYC's Excellence in Teaching Award, Diane Koenig was chosen as the Rock Valley College Faculty of the Year by her peers in 2005, and, in 2006, she was awarded the NISOD Teaching Excellence Award as well as the Illinois Mathematics Association of Community Colleges Award for Teaching Excellence. In addition to her teaching, Ms. Koenig has been an active member of the Illinois Mathematics Association of Community Colleges (IMACC). As a member, she has served on the board of directors, on a state-level task force rewriting the course outlines for the developmental mathematics courses, and as the association's newsletter editor.

Module 1: Operations with Integers

from Campus to Careers

Personal Financial Advisor

Personal financial advisors help people manage their money and teach them how to make their money grow. They offer advice on how to budget for monthly expenses, as well as how to save for retirement. A bachelor's degree in business, accounting, finance, economics, or statistics provides good preparation for the occupation. Strong communication and problem-solving skills are equally important to achieve success in this field.

In **Problem 104** of **Study Set 1.2,** you will see how a personal financial planner uses integers to determine whether a duplex rental unit would be a money-making investment for a client.

JOB TITLE: Personal Financial Advisor

EDUCATION: Must have at least a bachelor's degree. Some states require a certificate or license.

JOB OUTLOOK: Excellent—Jobs are projected to grow by 32% over the next decade.

ANNUAL EARNINGS: In 2010, average yearly earnings were $64,750.

FOR MORE INFORMATION: http://www.bls.gov/ooh/Business-and-Financial/Personal-financial-advisors.htm

Course Information Sheet

Overview

Module 1: Operations with Integers is one of the eight modules that make up the North Carolina Community College System Developmental Math Program. This program is for students who want to meet the prerequisites for the math requirements for their two year degree, or for those who are planning to transfer to a college or university. It is designed to allow you to complete the required developmental math courses at a pace that is appropriate to your needs and knowledge.

Placement

The diagnostic test that you took to enter the NCCCS Developmental Math Program has indentified your mathematical strengths and weaknesses. The test results that you received indicate which of the eight modules you are required to complete before you can enroll in more advanced mathematics courses, such as Precalculus and Statistics. It is important to note that any modules you are required to take must be taken in numerical order. For example, if the diagnostic test indicated that you need to take Modules 1 and 2, you must successfully complete Module 1 before you can register for Module 2.

Mastery

A core principle of the NCCCS Developmental Math Program is the concept of mastery of the material. To show mastery, students need to successfully complete all coursework in a module, as well as pass a final assessment exam.

Getting started

Starting a new course can be exciting, but it might also make you a bit nervous. In order to be successful, you need a plan. Here are some suggestions: Make time for the course, know what is expected, build a support system. You can begin to form your personal plan for success by answering questions on the next page.

1. What is your instructor's name? What is his/her phone number and email address?

Gary Comer 828 327 7000

gcomer@cvcc.edu

2. When and where does your class meet?

3. What are the days and times of your instructor's office hours? Where does he/she hold office hours?

4. Does your campus have a math tutoring center? If so, where is it located and what are its hours of operation? Is the tutoring free? Do you need your instructor to sign a form before you begin at the tutoring center?

5. What other ways are there for you to receive additional help with this module?

6. What are the names, phone numbers, and email addresses of three students in your class that you can contact for help if you have missed class, want to form a study group, or have questions regarding a homework assignment?

7. How many hours does your instructor feel you should expect to spend on this course each week?

8. Did you write down your WebAssign user id and password in a safe place where you can find it should you forget?

9. On what day and at what time is the final module assessment exam?

10. What percent correct is needed to pass the final module assessment exam? How many times can the final assessment exam be taken?

SECTION 1.1

Applications Introduction: Integers

In arithmetic, we add, subtract, multiply, and divide **whole numbers.**

Whole Numbers

The set of whole numbers is {0, 1, 2, 3, 4, 5, ...}. Read as, "the set containing zero, one, two, three, four, five, and so on."

Whole numbers, however, are not adequate for describing many real-life situations. For example, if you write a check for more than what's in your account, the account balance will be less than zero.

Numbers less than 0 are called **negative numbers,** and they can be represented on a **number line** by extending the line to the left. Negative numbers are always written with a **negative sign** $-$.

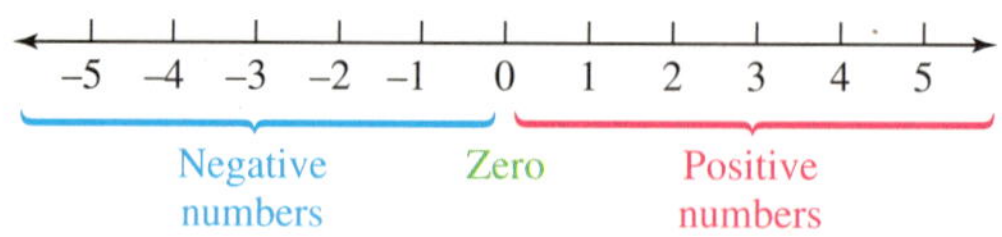

An important set of numbers called the *integers* was used to label the number line shown above.

Integers

The set of integers is {..., -4, -3, -2, -1, 0, 1, 2, 3, 4, ...}.

1. Does a *positive number* or a *negative number* best describe each situation below?

a. A 2-mile retreat by an army

b. A 15-pound weight gain

c. 10° above normal

d. 240 feet below sea level

e. 6 shots under par in golf

f. A $500 deposit

g. $15,000 in debt

h. A million dollar surplus

2. On the number line, what number is:

a. 4 units to the right of -3?

b. 6 units to the left of 1?

c. 1 unit to the left of -7?

d. 10 units to the right of -2?

e. 4 units to the right of -4?

f. 5 units to the left of 5?

SECTION 1.1
An Introduction to the Integers

Objectives

1 Define the set of integers.

2 Graph integers on a number line.

3 Use inequality symbols to compare integers.

4 Find the absolute value of an integer.

5 Find the opposite of an integer.

ARE YOU READY?

The following problems review some basic skills that are needed when working with integers.

1. Write the set of whole numbers.
2. Write the following numbers in order, from least to greatest: 7, 15, 0, 3, 26, 9
3. Draw a number line. Mark equally spaced points from 0 to 6.
4. Fill in the blank: When we speak of a submarine that is traveling underwater, we say it is _______ sea level.

The whole numbers are the numbers 0, 1, 2, 3, 4, 5, 6, 7, 8, 9, 10, ... We have seen that whole numbers can be used to describe many situations that arise in everyday life. However, we cannot use whole numbers to express temperatures below zero, the balance in a checking account that is overdrawn, or how far an object is below sea level. In this section, we will see how negative numbers can be used to describe these three situations as well as many others.

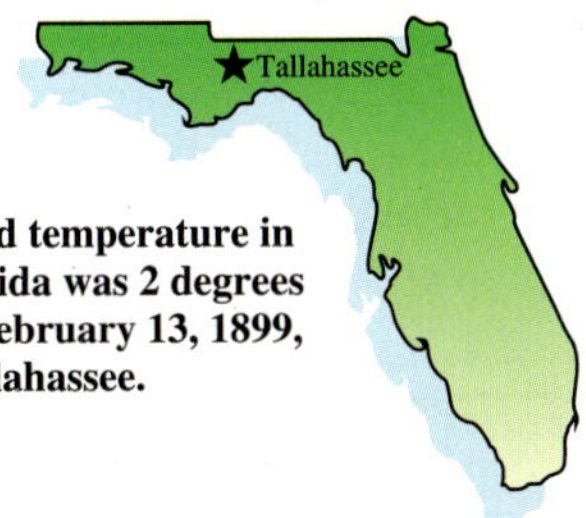

The record cold temperature in the state of Florida was 2 degrees *below* zero on February 13, 1899, in Tallahassee.

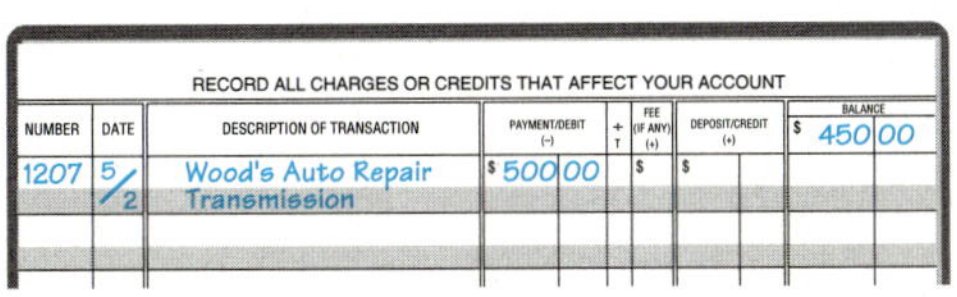

A check for $500 was written when there was only $450 in the account. The checking account is *overdrawn*.

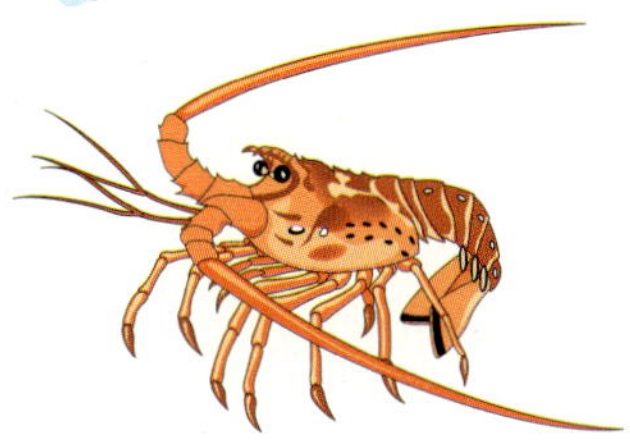

The American lobster is found off the Atlantic Coast from North Carolina to Newfoundland at depths as much as 600 feet *below* sea level.

1 Define the set of integers.

To describe a temperature of 2 degrees above zero, a balance of $50, or 600 feet above sea level, we can use numbers called **positive numbers.** All positive numbers are greater than 0, and we can write them with or without a **positive sign** +.

In words	**In symbols**	**Read as**
2 degrees above zero	+2 or 2	positive two
A balance of $50	+50 or 50	positive fifty
600 feet above sea level	+600 or 600	positive six hundred

To describe a temperature of 2 degrees below zero, \$50 overdrawn, or 600 feet below sea level, we need to use negative numbers. **Negative numbers** are numbers less than 0, and they are written using a **negative sign** $-$.

In words	**In symbols**	**Read as**
2 degrees below zero	-2	negative two
\$50 overdrawn	-50	negative fifty
600 feet below sea level	-600	negative six hundred

Together, positive and negative numbers are called **signed numbers.**

Positive and Negative Numbers

Positive numbers are greater than 0. **Negative numbers** are less than 0.

Caution! Zero is neither positive nor negative.

The collection of positive whole numbers, the negatives of the whole numbers, and 0 is called the set of **integers** (read as "in-ti-jers"). To write this set, we list its **members** (or **elements**) within **braces** { }.

The Set of Integers

$\{\ldots, -5, -4, -3, -2, -1, 0, 1, 2, 3, 4, 5, \ldots\}$

The three dots on the right indicate that the list continues forever—there is no largest integer. The three dots on the left indicate that the list continues forever—there is no smallest integer. The set of **positive integers** is $\{1, 2, 3, 4, 5, \ldots\}$ and the set of **negative integers** is $\{\ldots, -5, -4, -3, -2, -1\}$.

The Language of Mathematics Since every whole number is an integer, we say that the set of whole numbers is a **subset** of the integers.

The set of integers → $\{\ldots, -5, -4, -3, -2, -1, \underbrace{0, 1, 2, 3, 4, 5, \ldots}_{\text{The set of whole numbers}}\}$

2 Graph integers on a number line.

Negative numbers can be represented on a number line by extending the line to the left and drawing an arrowhead. Beginning at the origin (the 0 point), we move to the left, marking equally spaced points as shown below. As we move to the right on the number line, the values of the numbers increase. As we move to the left, the values of the numbers decrease.

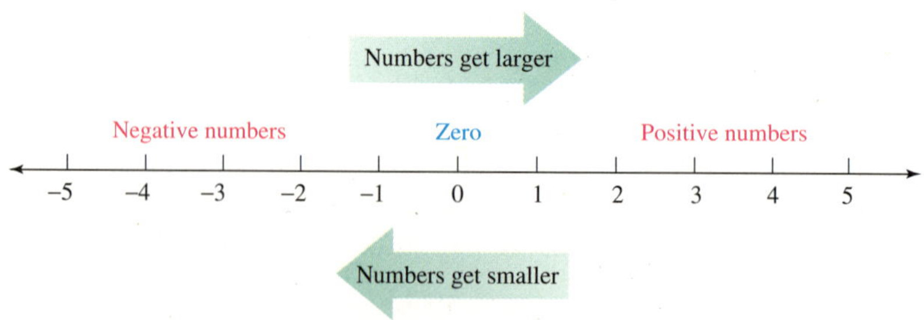

The thermometer shown on the next page is an example of a **vertical number line**. It is scaled in degrees and shows a temperature of $-10°$. The time line is an example of a **horizontal number line**. It is scaled in units of 500 years.

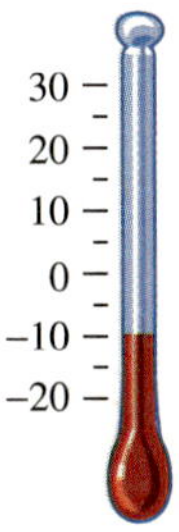

A vertical number line

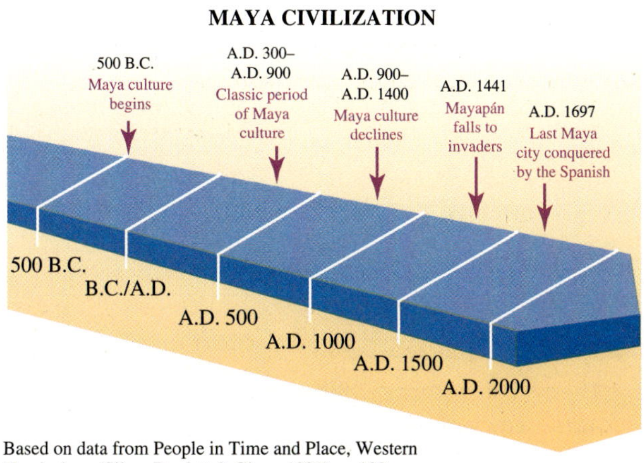

Based on data from People in Time and Place, Western Hemisphere (Silver Burdett & Ginn., 1991), p. 129

A horizontal number line

EXAMPLE 1 Graph −3, 2, −1, and 4 on a number line.

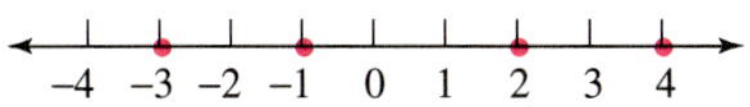

Strategy We will locate the position of each integer on the number line and draw a bold dot.

WHY To *graph a number* means to make a drawing that represents the number.

Solution

The position of each negative integer is to the left of 0. The position of each positive integer is to the right of 0.

Self Check 1

Graph −4, −2, 1, and 3 on a number line.

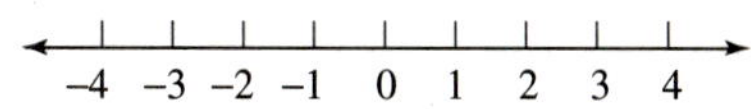

Now Try **Problem 27**

By extending the number line to include negative numbers, we can represent more situations using bar graphs and line graphs. For example, the following bar graph shows the net income of the Goodyear Tire and Rubber Company for the years 2004 through 2011. Since the net income in 2011 was positive \$321 million, the company made a profit that year. Since the net income in 2010 was −\$216 million, the company had a loss that year.

Goodyear Tire and Rubber company Net Income 2004-2011

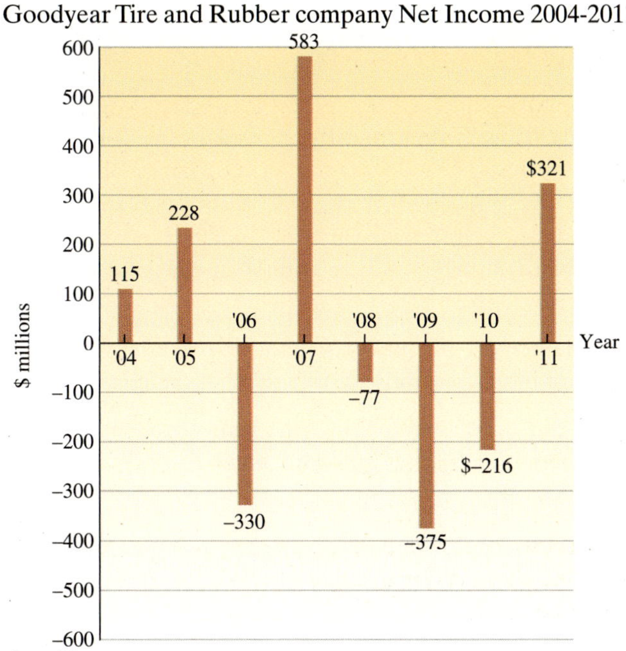

Source: YCharts

The Language of Mathematics *Net* refers to what remains after all the deductions (losses) have been accounted for. **Net income** is a term used in business that often is referred to as the *bottom line*. Net income indicates what a company has earned (or lost) in a given period of time (usually 1 year).

THINK IT THROUGH *Credit Card Debt*

"The most dangerous pitfall for many college students is the overuse of credit cards. Many banks do their best to entice new card holders with low or zero-interest cards."

Gary Schatsky, certified financial planner

Which numbers on the credit card statement below are actually debts and, therefore, could be represented using negative numbers?

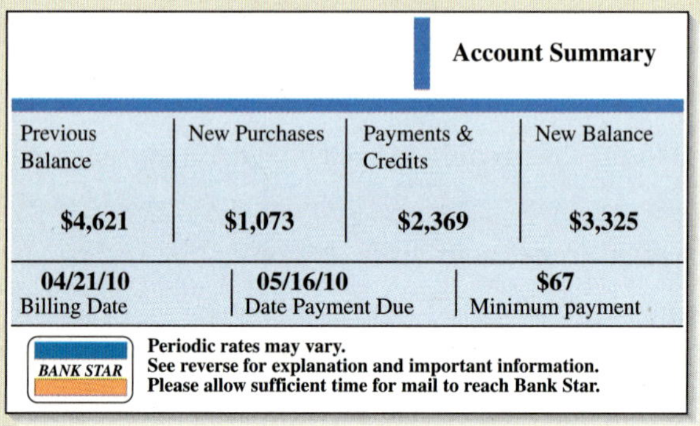

Account Summary

Previous Balance	New Purchases	Payments & Credits	New Balance
$4,621	$1,073	$2,369	$3,325

04/21/10 Billing Date	05/16/10 Date Payment Due	$67 Minimum payment

BANK STAR

Periodic rates may vary.
See reverse for explanation and important information.
Please allow sufficient time for mail to reach Bank Star.

3 Use inequality symbols to compare integers.

Recall that the symbol $<$ means "is less than" and that $>$ means "is greater than." The figure below shows the graph of the integers -2 and 1. Since -2 is to the left of 1 on the number line, $-2 < 1$. Since $-2 < 1$, it is also true that $1 > -2$.

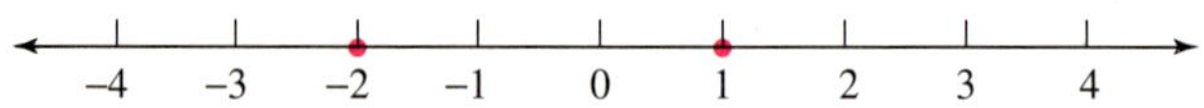

Self Check 2

Place an $<$ or an $>$ symbol in the box to make a true statement.

a. $6 \ \square\ -6$

b. $-11 \ \square\ -10$

Now Try **Problems 35 and 39**

EXAMPLE 2 Place an $<$ or an $>$ symbol in the box to make a true statement. **a.** $4 \ \square\ -5$ **b.** $-8 \ \square\ -7$

Strategy To pick the correct inequality symbol to place between the pair of numbers, we will determine the position of each number on the number line.

WHY For any two numbers on a number line, the number to the *left* is the smaller number and the number on the *right* is the larger number.

Solution

a. Since 4 is to the right of -5 on the number line, $4 > -5$.

b. Since -8 is to the left of -7 on the number line, $-8 < -7$.

The Language of Mathematics Because the symbol $<$ requires one number to be strictly less than another number and the symbol $>$ requires one number to be strictly greater than another number, mathematical statements involving the symbols $<$ and $>$ are called *strict inequalities.*

There are three other commonly used inequality symbols.

Inequality Symbols

$\neq$ means *is not equal to*

$\geq$ means *is greater than or equal to*

$\leq$ means *is less than or equal to*

$-5 \neq -2$ Read as "-5 is not equal to -2."

$-6 \leq 10$ Read as "-6 is less than or equal to 10." This statement is true, because $-6 < 10$.

$12 \leq 12$ Read as "12 is less than or equal to 12." This statement is true, because $12 = 12$.

$-15 \geq -17$ Read as "-15 is greater than or equal to -17." This statement is true, because $-15 > -17$.

$-20 \geq -20$ Read as "-20 is greater than or equal to -20." This statement is true, because $-20 = -20$.

EXAMPLE 3 Tell whether each statement is true or false.

a. $-9 \geq -9$ **b.** $-1 \leq -5$ **c.** $-27 \geq 6$ **d.** $-32 \leq -32$

Strategy We will determine if either the strict inequality or the equality that the symbols $\leq$ and $\geq$ allow is true.

WHY If either is true, then the given statement is true.

Solution

a. $-9 \geq -9$ This statement is true, because $-9 = -9$.

b. $-1 \leq -5$ This statement is false, because neither $-1 < -5$ nor $-1 = -5$ is true.

c. $-27 \geq 6$ This statement is false, because neither $-27 > 6$ nor $-27 = 6$ is true.

d. $-32 \leq -31$ This statement is true, because $-32 < -31$.

Self Check 3

Tell whether each statement is true or false.

a. $-17 \geq -15$

b. $-35 \leq -35$

c. $-2 \geq -2$

d. $-61 \leq -62$

Now Try **Problems 45 and 49**

4 Find the absolute value of an integer.

Using a number line, we can see that the numbers 3 and -3 are both a distance of 3 units away from 0, as shown below.

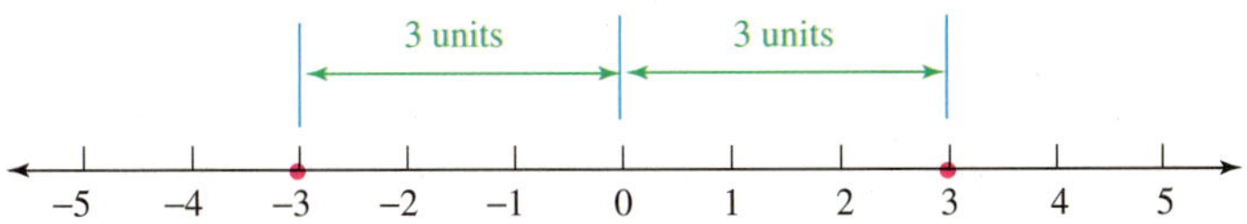

The **absolute value** of a number gives the distance between the number and 0 on the number line. To indicate absolute value, the number is inserted between two vertical bars, called the **absolute value symbol.** For example, we can write $|-3| = 3$. This is read as "The absolute value of negative 3 is 3," and it tells us that the distance between -3 and 0 on the number line is 3 units. From the figure, we also see that $|3| = 3$.

Absolute Value

The **absolute value** of a number is the distance on the number line between the number and 0.

Caution! Absolute value expresses distance. The absolute value of a number is always positive or 0. It is never negative.

Self Check 4

Find each absolute value:

a. $|-9|$

b. $|4|$

Now Try **Problems 51 and 53**

EXAMPLE 4 Find each absolute value: **a.** $|8|$ **b.** $|-5|$ **c.** $|0|$

Strategy We need to determine the distance that the number within the vertical absolute value bars is from 0 on a number line.

WHY The absolute value of a number is the distance between 0 and the number on a number line.

Solution

a. On the number line, the distance between 8 and 0 is 8. Therefore,

$$|8| = 8$$

b. On the number line, the distance between -5 and 0 is 5. Therefore,

$$|-5| = 5$$

c. On the number line, the distance between 0 and 0 is 0. Therefore,

$$|0| = 0$$

5 Find the opposite of an integer.

Opposites or Negatives

Two numbers that are the same distance from 0 on the number line, but on opposite sides of it, are called **opposites** or **negatives.**

The figure below shows that for each whole number on the number line, there is a corresponding whole number, called its *opposite,* to the left of 0. For example, we see that 3 and -3 are opposites, as are -5 and 5. Note that 0 is its own opposite.

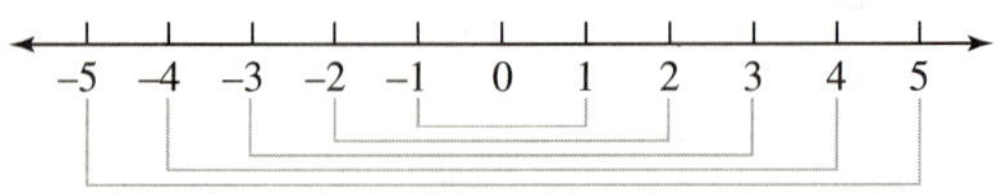

To write the opposite of a number, a $-$ symbol is used. For example, the opposite of 5 is -5 (read as "negative 5"). Parentheses are needed to express the opposite of a negative number. The opposite of -5 is written as $-(-5)$. Since 5 and -5 are the same distance from 0, the opposite of -5 is 5. Therefore, $-(-5) = 5$. This illustrates the following rule.

The Opposite of the Opposite Rule

The opposite of the opposite (or negative) of a number is that number.

Number	Opposite	
57	-57	Read as "negative fifty-seven."
-8	$-(-8) = 8$	Read as "the opposite of negative eight is eight."
0	$-0 = 0$	Read as "the opposite of 0 is 0."

The concept of opposite can also be applied to an absolute value. For example, the opposite of the absolute value of -8 can be written as $-|-8|$. Think of this as a two-step process, where the absolute value symbol serves as a grouping symbol. Find the absolute value first, and then attach a $-$ sign to that result.

First, find the absolute value.

$$-|-8| = -8$$

Read as "the opposite of the absolute value of negative eight is negative eight."

Then attach a $-$ sign.

EXAMPLE 5 Simplify each expression: **a.** $-(-44)$ **b.** $-|11|$ **c.** $-|-225|$

Strategy We will find the opposite of each number.

WHY In each case, the $-$ symbol written outside the grouping symbols means "the opposite of."

Solution

a. $-(-44)$ means the opposite of -44. Since the opposite of -44 is 44, we write

$$-(-44) = 44$$

b. $-|11|$ means the opposite of the absolute value of 11. Since $|11| = 11$, and the opposite of 11 is -11, we write

$$-|11| = -11$$

c. $-|-225|$ means the opposite of the absolute value of -225. Since $|-225| = 225$, and the opposite of 225 is -225, we write

$$-|-225| = -225$$

Self Check 5

Simplify each expression:

a. $-(-1)$

b. $-|4|$

c. $-|-99|$

Now Try **Problems 59, 69, and 71**

The $-$ symbol is used to indicate a negative number, the opposite of a number, and the operation of subtraction. The key to reading the $-$ symbol correctly is to examine the context in which it is used.

Reading the – Symbol

-12	Negative twelve	A $-$ symbol directly in front of a number is read as "negative."
$-(-12)$	The opposite of negative twelve	The first $-$ symbol is read as "the opposite of" and the second as "negative."
$12 - 5$	Twelve minus five	Notice the space used before and after the $-$ symbol. This indicates subtraction and is read as "minus."

ANSWERS TO SELF CHECKS

1. 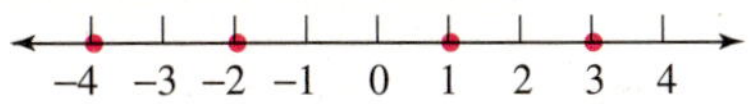

2. a. $>$ **b.** $<$
3. a. false **b.** true **c.** true **d.** false **4. a.** 9 **b.** 4 **5. a.** 1 **b.** -4 **c.** -99

SECTION 1.1 STUDY SET

VOCABULARY

Fill in the blanks.

1. ________ numbers are greater than 0 and ________ numbers are less than 0.
2. Positive and negative numbers are called ______ numbers.
3. 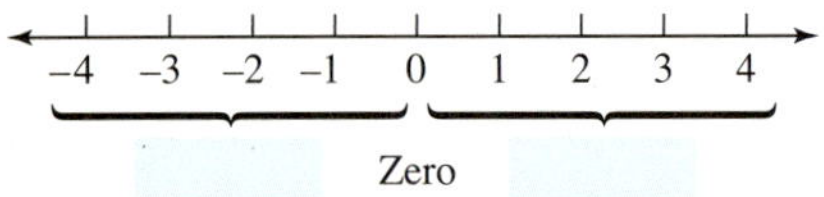

4. $\{\ldots, -5, -4, -3, -2, -1, 0, 1, 2, 3, 4, 5, \ldots\}$ is called the set of ________.
5. To _______ an integer means to locate it on the number line and highlight it with a dot.
6. The symbols $>$ and $<$ are called __________ symbols.
7. The ________ ________ of a number is the distance between the number and 0 on the number line.
8. Two numbers that are the same distance from 0 on the number line, but on opposite sides of it, are called ________.

CONCEPTS

9. Represent each of these situations using a signed number.
   - **a.** \$225 overdrawn
   - **b.** A 15-pound weight gain
   - **c.** 10 seconds before liftoff
   - **d.** 3 degrees below normal
   - **e.** 10° above zero
   - **f.** A deficit of \$12,000
   - **g.** A 1-mile retreat by an army
   - **h.** An elevation of 15,257 feet
10. Represent each of these situations using a signed number, and then describe its opposite in words.
    - **a.** A trade surplus of \$3 million
    - **b.** 8 steps backward
    - **c.** A bacteria count 70 more than the standard
    - **d.** A profit of \$67
    - **e.** A business \$1 million in the "black"
    - **f.** 20 units over their quota
    - **g.** 312 feet below the surface
    - **h.** 16° below zero
11. Copy the number line below. Then represent each quantity as an integer on the number line.

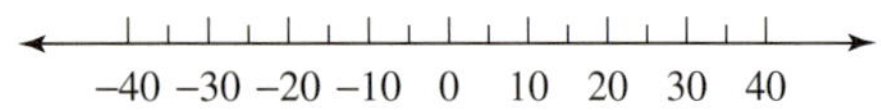

    - **a.** 10 degrees above zero
    - **b.** A building foundation 15 inches below grade
    - **c.** A 35-centimeter rise
    - **d.** 40 feet below sea level
12. **a.** If a number is less than 0, what type of number must it be?
    b. If a number is greater than 0, what type of number must it be?
13. On the number line, what number is
    - **a.** 3 units to the right of -7?
    - **b.** 4 units to the left of 2?
14. Name two numbers on the number line that are a distance of
    - **a.** 5 away from -3.
    - **b.** 4 away from 3.
15. **a.** Which number is closer to -3 on the number line: 2 or -7?
    b. Which number is farther from 1 on the number line: -5 or 8?
16. Is there a number that is both greater than 10 and less than 10 at the same time?
17. **a.** Express the fact $-12 < 15$ using an $>$ symbol.
    b. Express the fact $-4 > -5$ using an $<$ symbol.
18. Fill in the blank: The opposite of the ________ of a number is that number.
19. Complete the table by finding the opposite and the absolute value of the given numbers.

Number	Opposite	Absolute value
−25		
39		
0		

20. Is the absolute value of a number always positive?

NOTATION

21. Translate each phrase to mathematical symbols.

a. The opposite of negative eight

b. The absolute value of negative eight

c. Eight minus eight

d. The opposite of the absolute value of negative eight

22. a. Write the set of integers.

b. Write the set of positive integers.

c. Write the set of negative integers.

23. Fill in the blanks.

a. We read $\geq$ as "is _______ than or _______ to."

b. We read $\leq$ as "is _____ than or _______ to."

24. $|-15|$ is read "the _______ _____ of -15."

25. The symbols { } are called ______.

26. Which of the following expressions contains a minus sign?

$15 - 8 \qquad -(-15) \qquad -15$

GUIDED PRACTICE

Graph the following numbers on a number line. See Example 1.

27. $-3, 4, 3, 0, -1$

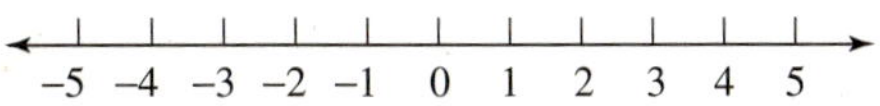

28. $2, -4, 5, 1, -1$

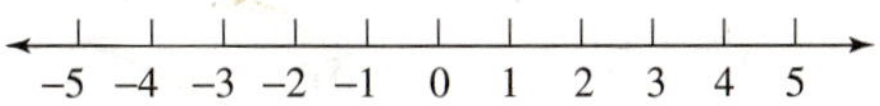

29. The integers that are less than 3 but greater than -5

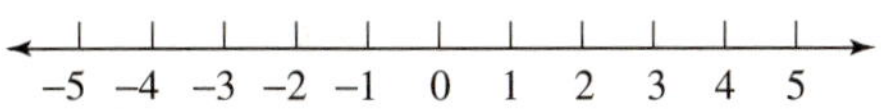

30. The integers that are less than 4 but greater than -3

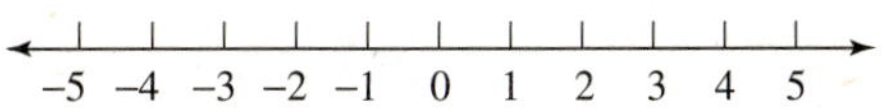

31. The opposite of -3, the opposite of 5, and the absolute value of -2

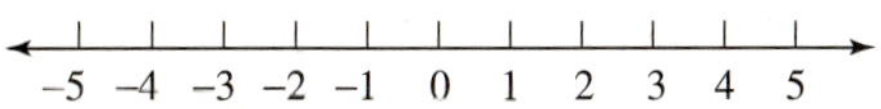

32. The absolute value of 3, the opposite of 3, and the number that is 1 less than -3

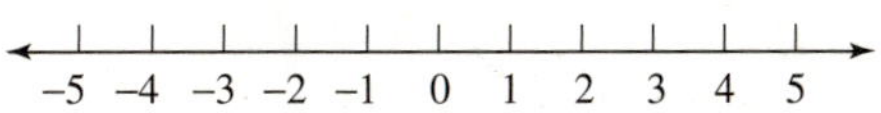

33. 2 more than 0, 4 less than 0, 2 more than negative 5, and 5 less than 4

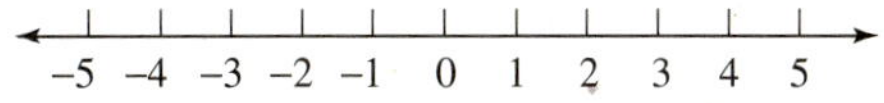

34. 4 less than 0, 1 more than 0, 2 less than -2, and 6 more than -4

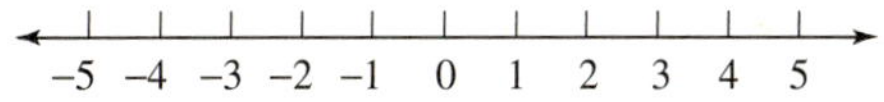

Place an < or an > symbol in the box to make a true statement. See Example 2.

35. -5 □ 5 **36.** 0 □ -1

37. -12 □ -6 **38.** -7 □ -6

39. -10 □ -17 **40.** -11 □ -20

41. -325 □ -532 **42.** -401 □ -104

Tell whether each statement is true or false. See Example 3.

43. $-15 \leq -14$ **44.** $-77 \leq -76$

45. $210 \geq 210$ **46.** $37 \geq 37$

47. $-1{,}255 \geq -1{,}254$ **48.** $-6{,}546 \geq -6{,}465$

49. $0 \leq -8$ **50.** $-6 \leq -6$

Find each absolute value. See Example 4.

51. $|9|$ **52.** $|12|$

53. $|-8|$ **54.** $|-1|$

55. $|-14|$ **56.** $|-85|$

57. $|180|$ **58.** $|371|$

Simplify each expression. See Example 5.

59. $-(-11)$ **60.** $-(-1)$

61. $-(-4)$ **62.** $-(-9)$

63. $-(-102)$ **64.** $-(-295)$

65. $-(-561)$ **66.** $-(-703)$

67. $-|20|$ **68.** $-|143|$

69. $-|6|$ **70.** $-|0|$

71. $-|-253|$ **72.** $-|-11|$

73. $-|-0|$ **74.** $-|97|$

TRY IT YOURSELF

Place an < or an > symbol in the box to make a true statement.

75. $|-12|$ □ $-(-7)$ **76.** $|-50|$ □ $-(-40)$

77. $-|-71|$ □ $-|-65|$ **78.** $-|-163|$ □ $-|-150|$

79. $-(-343)$ □ $-(-161)$ **80.** $-(-999)$ □ $-(-998)$

81. $-|-30|$ □ $-|-(-8)|$ **82.** $-|-100|$ □ $-|-(-88)|$

Write the following integers in order, from least to greatest.

83. 82, −52, 52, −22, 12, −12

84. 49, −9, 19, −39, 89, −49

Fill in the blanks to continue each pattern.

85. 5, 3, 1, −1, ___, ___, ___, . . .

86. 4, 2, 0, −2, ___, ___, ___, . . .

87. Refer to the number line below. Use an inequality symbol, $<$ or $>$, to make each statement true.

a. a ___ b **b.** b ___ a

c. b ___ 0 and a ___ 0 **d.** $|a|$ ___ $|b|$

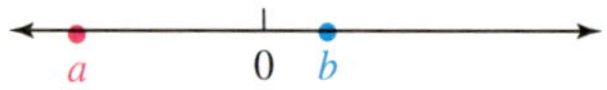

88. Let a represent a number greater than zero and b represent a number less than 0. Determine whether each statement is *sometimes true*, *always true*, or *never true*. Explain your reasoning.

a. $a > -a$

b. $b > -b$

c. $|a| > |b|$

d. $|b| > 0$

89. How many integers have an absolute value that is less than 1,000?

90. What is the opposite of the opposite of any integer?

APPLICATIONS

91. HORSE RACING In the 1973 Belmont Stakes, *Secretariat* won by 31 lengths over second place finisher, *Twice a Prince.* Some experts call it the greatest performance by a thoroughbred in the history of racing. Express the position of *Twice a Prince* compared to *Secretariat* as a signed number. (Source: ezinearticles.com)

© Bettmann/Corbis

92. NASCAR In the NASCAR driver standings, negative numbers are used to tell how many points behind the leader a given driver is. Jimmie Johnson was the leading driver in 2008. The other drivers in the top ten were Greg Biffle (−217), Clint Bowyer (−303), Jeff Burton (−349), Kyle Busch (−498), Carl Edwards (−69), Jeff Gordon (−368), Denny Hamlin (−470), Kevin Harvick (−276), and Tony Stewart (−482). Use this information to rank the drivers in the table below.

AP Images

2008 NASCAR Final Driver Standings

Rank	Driver	Points behind leader
1	Jimmie Johnson	Leader
2		
3		
4		
5		
6		
7		
8		
9		
10		

(Source: NASCAR.com)

93. FREE FALL A boy launches a water balloon from the top of a building, as shown below. At that instant, his friend starts a stopwatch and keeps track of the time as the balloon sails above the building and then falls to the ground. Use the number line to estimate the position of the balloon at each time listed in the table on the next page.

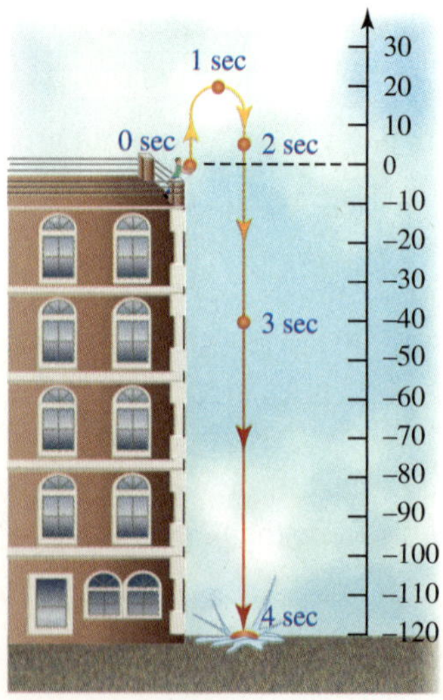

Time	Position of balloon
0 sec	
1 sec	
2 sec	
3 sec	
4 sec	

94. CARNIVAL GAMES At a carnival shooting gallery, players aim at moving ducks. The path of one duck is shown, along with the time it takes the duck to reach certain positions on the gallery wall. Use the number line to estimate the position of the duck at each time listed in the table below.

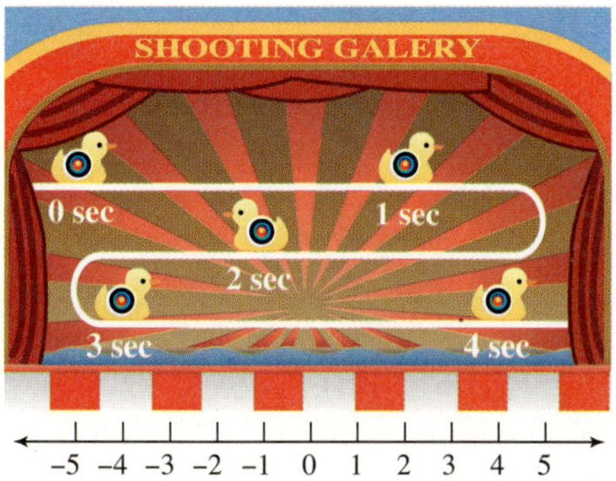

Time	Position of duck
0 sec	
1 sec	
2 sec	
3 sec	
4 sec	

95. TECHNOLOGY The readout from a testing device is shown. Use the number line to find the height of each of the peaks and the depth of each of the valleys.

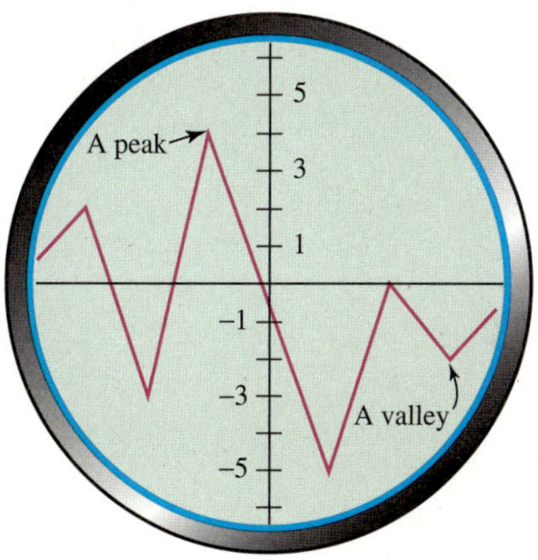

96. FLOODING A week of daily reports listing the height of a river in comparison to flood stage is given in the table. Complete the bar graph shown below.

Flood Stage Report

Sun.	2 ft below
Mon.	3 ft over
Tue.	4 ft over
Wed.	2 ft over
Thu.	1 ft below
Fri.	3 ft below
Sat.	4 ft below

97. GOLF In golf, *par* is the standard number of strokes considered necessary on a given hole. A score of −2 indicates that a golfer used 2 strokes less than par. A score of +2 means 2 more strokes than par were used. In the graph below, each golf ball represents the score of a professional golfer on the 16th hole of a certain course.

a. What score was shot most often on this hole?

b. What was the best score on this hole?

c. Explain why this hole appears to be too easy for a professional golfer.

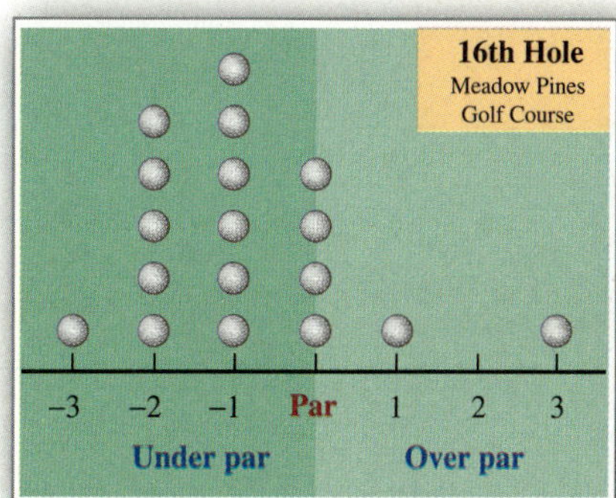

98. PAYCHECKS Examine the items listed on the paycheck stub on the next page. Then write two columns on your paper—one headed "positive" and the other "negative." List each item under the proper heading.

Tom Dryden	**Dec. 12**	Christmas bonus	$100
Gross pay	$2,000	**Reductions**	
Overtime	$300	Retirement	$200
Deductions		**Taxes**	
Union dues	$30	Federal withholding	$160
U.S. Bonds	$100	State withholding	$35

99. WEATHER MAPS The illustration shows the predicted Fahrenheit temperatures for a day in mid-January.

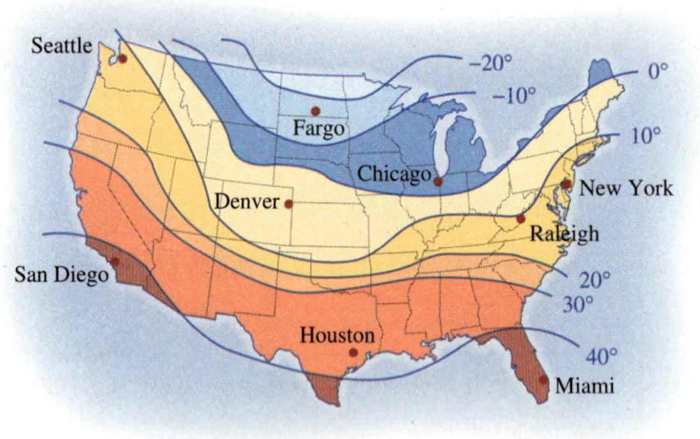

a. What is the temperature range for the region including Raleigh, North Carolina?

b. What is the temperature range for the region including Fargo, North Dakota?

c. According to the prediction, what is the warmest it should get in Houston?

d. According to this prediction, what is the coldest it should get in Seattle?

100. INTERNET COMPANIES The graph below shows the net income of Amazon.com for the years 1998–2011. (Source: Morningstar)

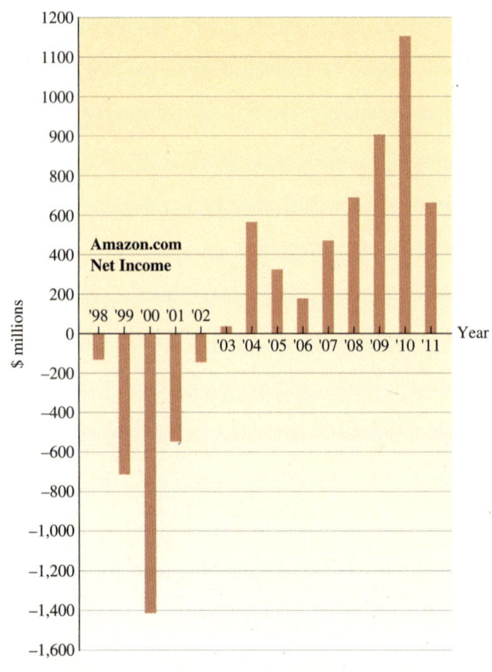

a. In what years did Amazon suffer a loss? Estimate each loss.

b. In what year did Amazon first turn a profit? Estimate it.

c. In what year did Amazon have the greatest profit? Estimate it.

101. HISTORY Number lines can be used to display historical data. Some important world events are shown on the time line below.

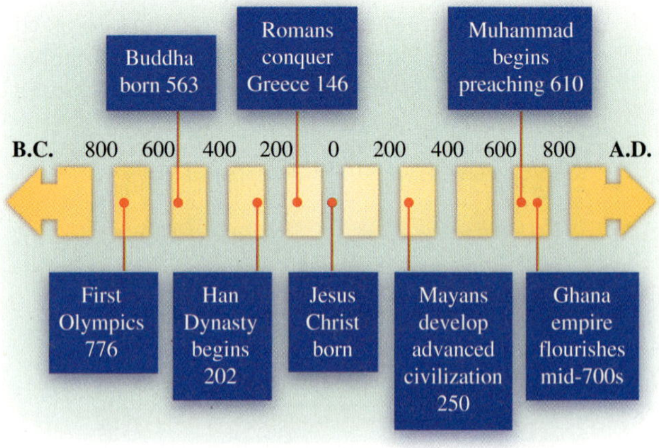

a. What basic unit is used to scale this time line?

b. What can be thought of as positive numbers?

c. What can be thought of as negative numbers?

d. What important event distinguishes the positive from the negative numbers?

102. ASTRONOMY Astronomers use an inverted vertical number line called the *apparent magnitude scale* to denote the brightness of objects in the sky. The brighter an object appears to an observer on Earth, the more negative is its apparent magnitude. Graph each of the following on the scale to the right.

- Visual limit of binoculars +10
- Visual limit of large telescope +20
- Visual limit of naked eye +6
- Full moon −12
- Pluto +15
- Sirius (a bright star) −2
- Sun −26
- Venus −4

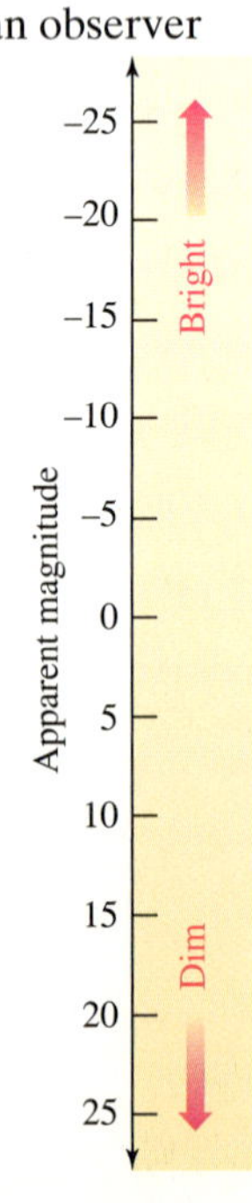

103. LINE GRAPHS Each thermometer in the illustration gives the daily high temperature in degrees Fahrenheit. Use the data to complete the line graph below.

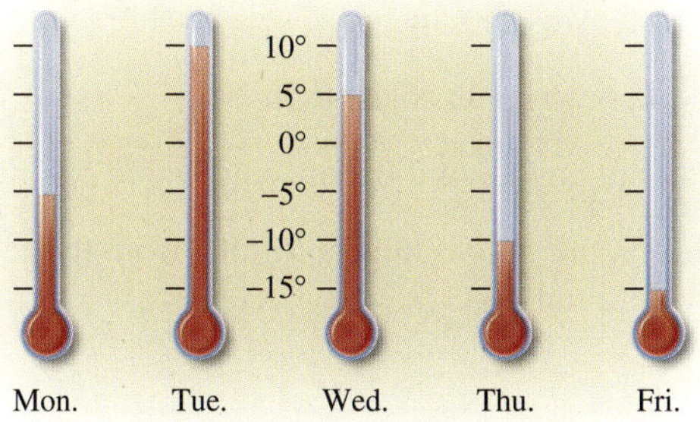

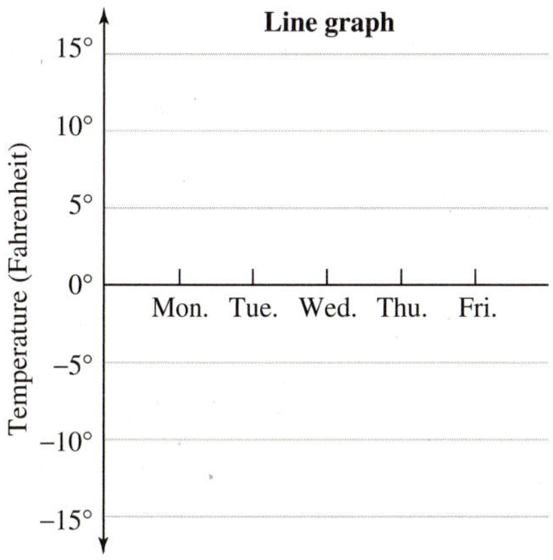

104. GARDENING The illustration shows the depths at which the bottoms of various types of flower bulbs should be planted. (The symbol ″ represents inches.)

a. At what depth should a tulip bulb be planted?

b. How much deeper are hyacinth bulbs planted than gladiolus bulbs?

c. Which bulb must be planted the deepest? How deep?

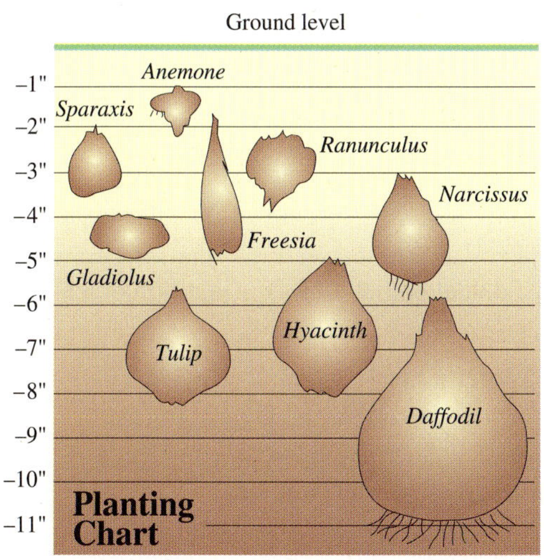

105. IPHONES You can get a more accurate reading of an iPhone's signal strength by dialing *3001#12345#*. Field test mode is then activated and the standard signal strength bars (in the upper left corner of the display) are replaced by a negative number. The closer the negative number is to zero, the stronger the signal. Which iPhone shown below is receiving the strongest signal?

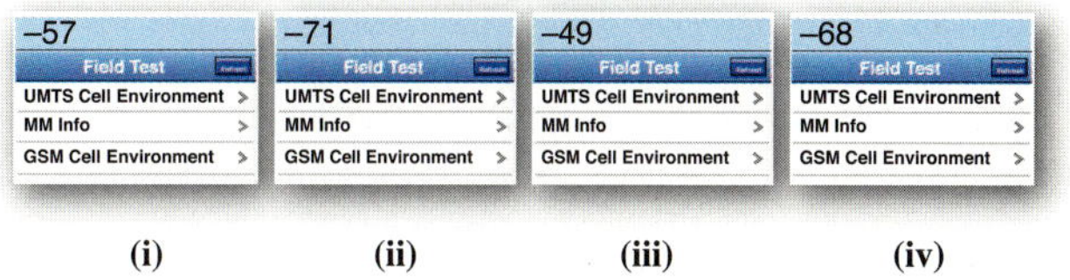

106. U.S. BUDGET A budget *deficit* is a negative number that indicates the government spent more money than it took in that year. A budget *surplus* is a positive number that indicates the government took in more money than it spent that year.

a. Refer to the graph below that shows the U.S. Federal Budget Deficit/Surplus for the years 1980 through 2009. For how many of those years was there a budget surplus?

b. Consider the years in which there was a budget deficit. For how many of those years was it smaller than \$300 billion?

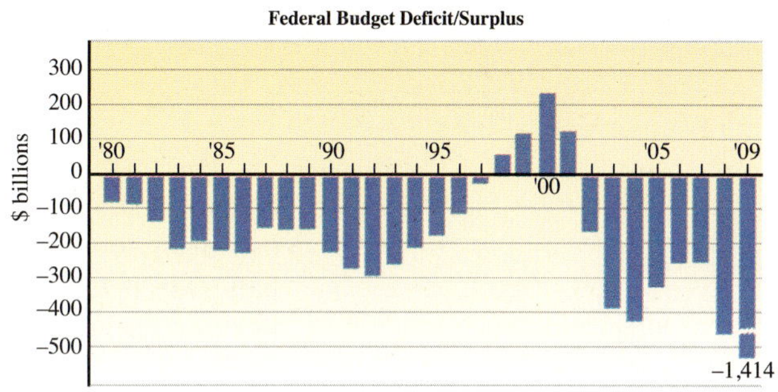

Source: U.S. Bureau of the Census

WRITING

107. Explain the concept of *the opposite of a number.*

108. What real-life situation do you think gave rise to the concept of a negative number?

109. Explain why the absolute value of a number is never negative.

110. Give an example of the use of the number line that you have seen in another course.

111. DIVING Divers use the terms *positive buoyancy, neutral buoyancy,* and *negative buoyancy* as shown. What do you think each of these terms means?

112. GEOGRAPHY Much of the Netherlands is low-lying, with half of the country below sea level. Explain why it is not under water.

113. Suppose integer A is greater than integer B. Is the opposite of integer A greater than integer B? Explain why or why not. Use an example.

114. Explain why -11 is less than -10.

115. Explain how to find the absolute value of a number.

116. How is the set of integers formed from the set of whole numbers?

SECTION 1.2
Applications Introduction: Adding Integers Using Models

Models are helpful when learning about integer addition. They can give you a fundamental understanding of the concept so that you aren't simply memorizing a set of rules that can be forgotten in a few days.

Heaps and holes model

One way in which integer addition is modeled is using **heaps** and **holes**. With this method, we represent positive 1 with one heap (or pile) of sand on a level beach. A hole of the same size dug into the beach represents -1.

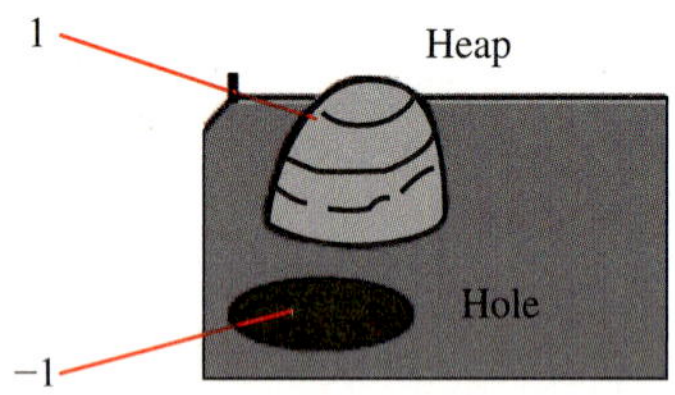

Notice that one heap perfectly fills in one hole. The result is 0.

1 heap + 1 hole = 0

We can represent any positive integer using a collection of heaps and any negative integer can be represented by a collection of holes. To make the drawing easier, simple half-ovals above a horizontal line (heaps) are used to represent positive integers and half-ovals below a horizontal line (holes) are used to represent negative integers. The illustrations below show how to represent positive 5 and -3.

5 heaps

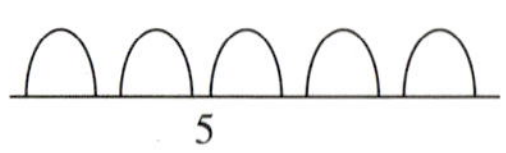

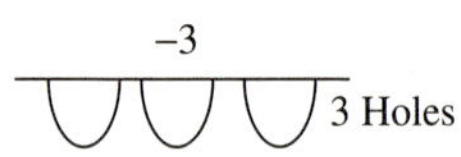

To perform addition using heaps and holes, we draw the integers on a single model. For example, the heap/hole model below shows the addition of $5 + (-3)$.

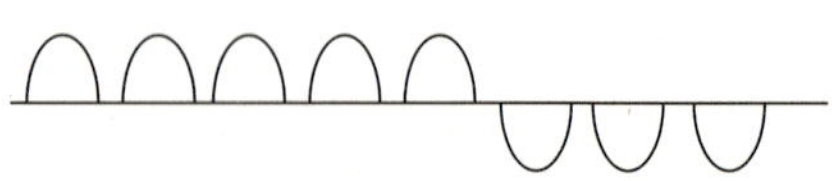

$5 + (-3) = ?$

If we slide three heaps over three holes to fill them, three 0s are formed. We are left with 2 heaps, which represent the result of the addition.

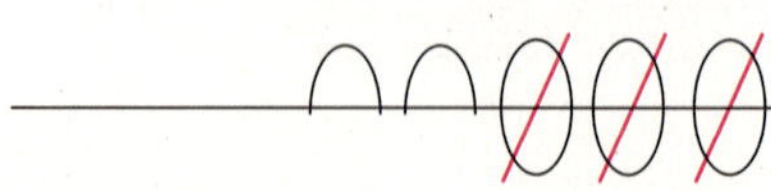

$5 + (-3) = 2$

1. Complete each number sentence that the heaps and holes model illustrates.

 a. 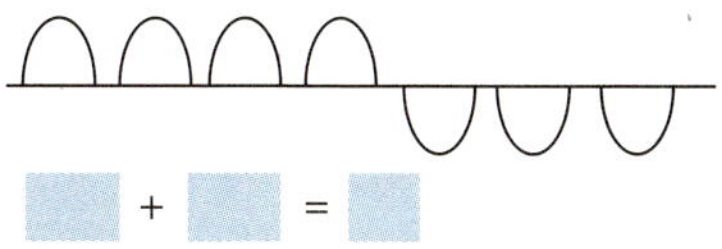

 b. 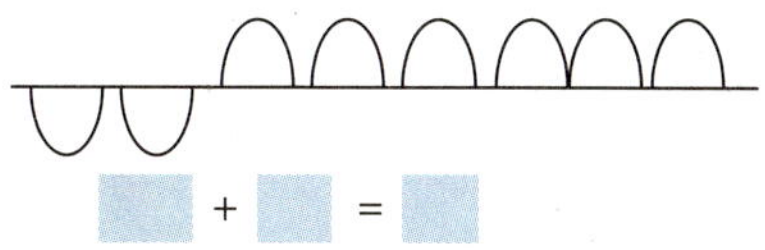

2. Draw a heaps and holes model to find each result.

 a. $6 + (-2)$ b. $4 + (-7)$

 c. $-5 + 5$ d. $-3 + (-8)$

 e. $-90 + 30$ (*Hint:* Let one heap represent 10 and one hole represent -10)

 f. $700 + (-200)$ (*Hint:* Let one heap represent 100 and one hole represent -100)

 g. $-1 + 4 + (-5)$

3. a. Draw a heaps and holes model for $8 + (-6)$ and find the result.

 b. Now switch the order of the numbers being added. Draw a heaps and holes model for $-6 + 8$ and find the result. Do you get the same answer as part a?

4. The correct answer to problem 3b is, "Yes, we get the same answer." Problem 3 illustrates an important fact called the **commutative property of addition:** The order in which integers are added does not change their sum. Determine whether the order in which the following actions are taken always produces the same result.

 a. Put on your left sock and put on your right sock

 b. Put frozen food in a microwave oven and press the "Cook" button

 c. Wash your clothes and dry them

 d. Pour a cup of coffee and add some cream

 e. Wash your face and comb your hair

 f. Jump out of an airplane and pull the parachute ripcord

Colored tiles model

Another way to model integer addition is using **colored tiles.** With this model, we use a blue tile to represent positive 1 and a red tile to represent -1.

We can represent any positive integer using a collection of blue tiles and any negative integer can be represented by a collection of red tiles. The illustrations below show how to represent positive 4 and -7.

Together, a blue tile and a red tile are called a **zero pair** because their sum is zero: $1 + (-1) = 0$ and $-1 + 1 = 0$.

When adding integers using colored tiles, we look for zero pairs. For example, to find $-7 + 4$, notice that we can form four zero pairs, as shown below. We are left with 3 red tiles, which represent the result of the addition.

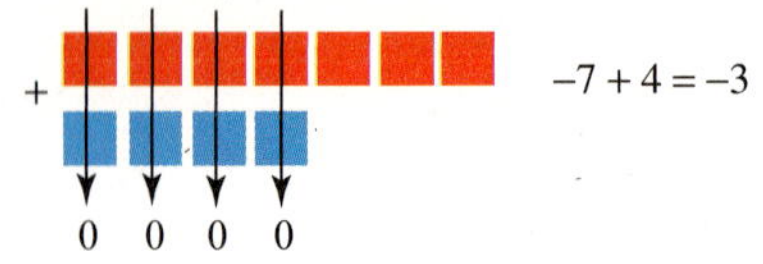

5. Complete each number sentence that the colored tile model illustrates.

a.

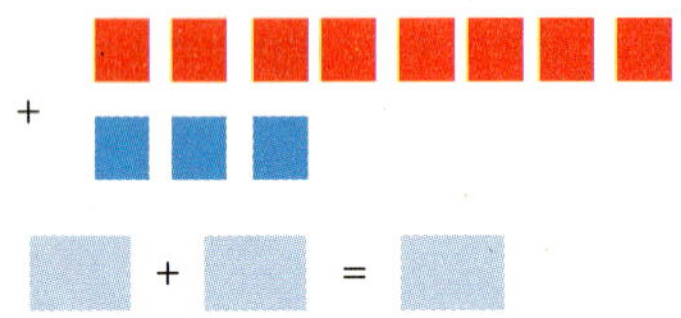

b.

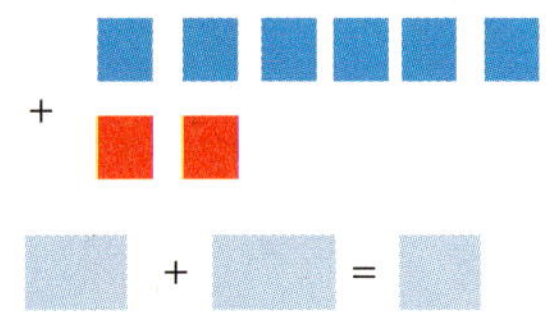

c. How many zero pairs did you find in part a?

d. How many zero pairs did you find in part b?

6. Draw a colored tile model to find each result.

a. $7 + (-2)$

b. $6 + (-8)$

c. $-9 + 4$

d. $-3 + 3$

e. $-50 + 10$ (*Hint:* Let one blue tile represent 10 and one red tile represent -10)

f. $900 + (-700)$ (*Hint:* Let one blue tile represent 100 and one red tile represent -100)

g. $-4 + (-2) + 7$

Number line model

With this model, we show addition of integers by thinking in terms of *distance* and *direction*.

7. To become familiar with movement on a number line, answer the following questions.

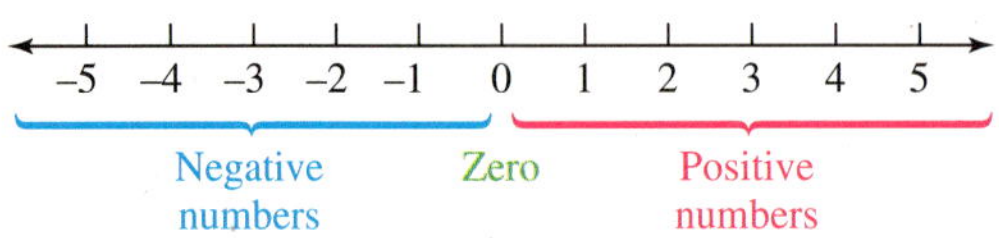

a. Begin at 0 on the number line and move 3 units to the right followed by a move of 4 units to the right. Is the result in the positive or the negative region of the number line?

b. Begin at 0 on the number line and move 2 units to the left followed by a move of 1 unit to the left. Is the result in the positive or the negative region of the number line?

c. Begin at 0 on the number line and move 4 units to the right followed by a move of 5 units to the left. Is the result in the positive or the negative region of the number line?

d. Begin at 0 on the number line and move 1 unit to the left followed by a move of 3 units to the right. Is the result in the positive or the negative region of the number line?

Integer addition can be shown on a number line using **arrows.** The arrows display distance and a direction. A positive integer is represented by an arrow pointing to the right, and a negative integer by an arrow pointing to the left.

The addition process has two steps. We begin at 0 on the number line and draw an arrow of the proper length in the proper direction to represent the first integer. From the *tip* of that arrow, we draw another arrow of the proper length in the proper direction to represent the second integer. The tip of the second arrow points to the result of the addition.

8. The following number line models illustrate integer addition. Fill in the blank.

a.

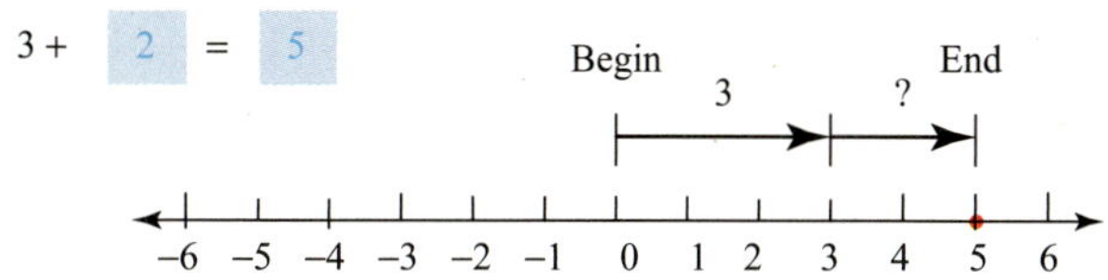

b.

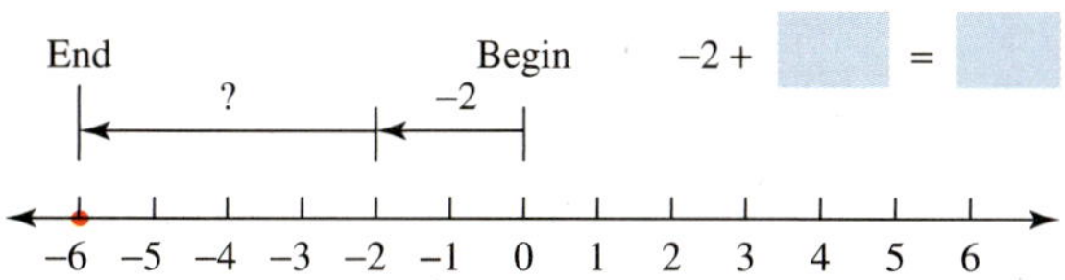

c.

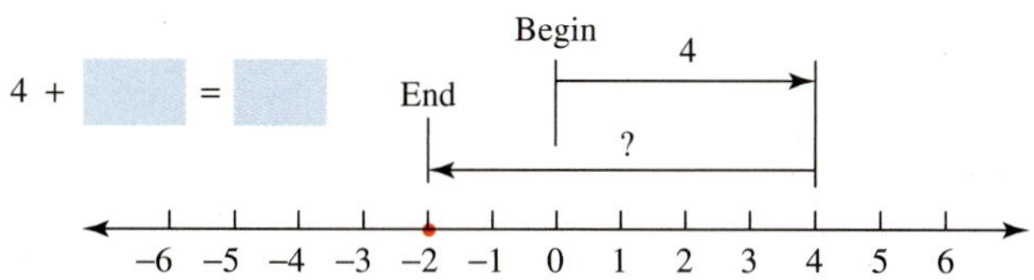

d.

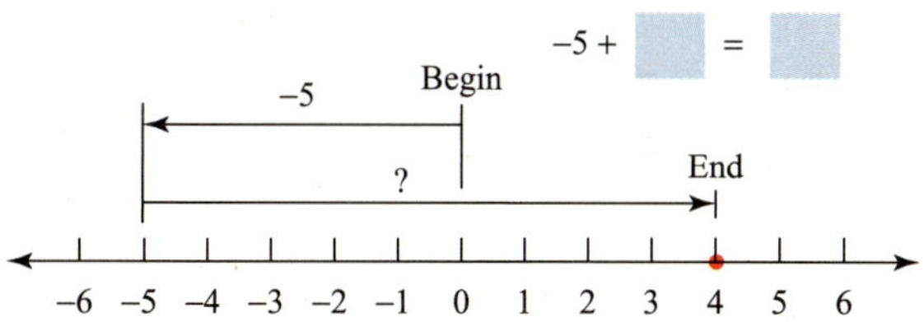

Win/lose money model

In the news, we often hear of huge lottery winnings and devastating financial losses. This contrast suggests another way to model integer addition. With the **winning and losing money model,** we let positive integers represent winning money and negative integers represent losing money.

9. Game shows. Determine whether the score of a contestant on the game show Jeopardy! will be positive or negative after two questions if she answers in the following way.

a. Correctly answers the first question to win \$400, but incorrectly answers the second question to lose \$100

b. Correctly answers the first question to win \$100, but incorrectly answers the second question to lose \$200

c. Incorrectly answers the first question to lose \$300, and incorrectly answers the second question to lose \$400

d. Incorrectly answers the first question to lose \$100, but correctly answers the second question to win \$500

10. Gambling. For each pair of outcomes, tell whether the gambler would end up *ahead* or *behind*, and by how much.

a. Won \$50 and then lost \$30

b. Lost \$50 and then won \$30

c. Won \$50 and then won \$30

d. Lost \$30 and then lost \$50

e. Won \$30 and then lost \$50

f. Won \$30 and then won \$50

g. Lost \$50 and then lost \$30

SECTION 1.2
Adding Integers

Objectives

1. Model integer addition.
2. Add two integers that have the same sign.
3. Add two integers that have different signs.
4. Perform several additions to evaluate expressions.
5. Use properties of addition.
6. Identify opposites (additive inverses) when adding integers.
7. Solve application problems by adding integers.

ARE YOU READY?

The following problems review some basic concepts that are important when adding positive and negative integers.

1. Find $|3|$ and $|-5|$. Which number, 3 or -5, has the larger absolute value?

2. What is the opposite of -11?

3. Subtract: $710 - 89$

4. If you begin at 0 on a number line and move 5 units in the positive direction, and then 4 more units in the positive direction, where will you end up on the number line?

An amazing change in temperature occurred in 1943 in Spearfish, South Dakota. On January 22, at 7:30 A.M., the temperature was -4 degrees Fahrenheit. Strong warming winds suddenly kicked up and, in just 2 minutes, the temperature rose 49 degrees! To calculate the temperature at 7:32 A.M., we need to add 49 to -4.

$$-4 + 49$$

To perform this addition, we must know how to add positive and negative integers. In this section, we develop rules to help us make such calculations.

1 Model integer addition.

Models can be helpful in developing a deeper understanding of the rules for adding integers. Four of the most popular models are used to illustrate the addition problem $4 + (-3)$ below. See pages 18 through 21 of the Applications Introduction for a more detailed explanation of each model.

- ***Heaps and holes model:*** We represent positive 1 with one heap (or pile) of sand on a level beach. A hole of the same size dug into the beach represents -1. To make the drawing easier, simple half-ovals above a horizontal line (heaps) are used to represent positive integers and half-ovals below a horizontal line (holes) are used to represent negative integers. In this model use the fact that 1 heap perfectly fills in 1 hole. For each pairing the result is 0.

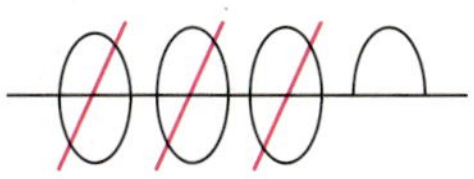

- ***Colored tiles model:*** We use a blue tile to represent positive 1 and a red tile to represent -1. Together, a blue tile and a red tile are called a **zero pair** because their sum is zero: $1 + (-1) = 0$ and $-1 + 1 = 0$.

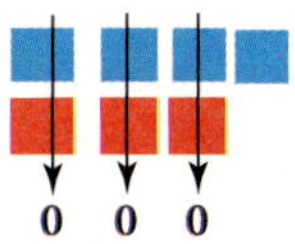

- ***Number line model:*** We draw an **arrow to the right** of a given length to represent a positive number and an **arrow to the left** of a given length to represent a negative number.

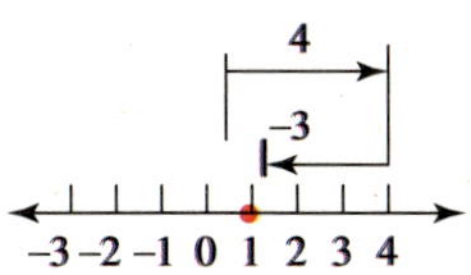

- ***Winning/losing money model:*** We think of a positive number as an amount of money that is won and a negative number as an amount of money that is lost.

2 Add two integers that have the same sign.

When adding integers, there are two cases to consider: Either both integers have the same sign, or they have different signs. To explore the first possibility, where both integers are positive, let's look at the problem $4 + 3$. We can use a number line model to find the result.

On a number line, we begin at 0 and draw an arrow 4 units long that points to the right. It represents positive 4. From the tip of that arrow, we draw a second arrow, 3 units long, that points to the right. It represents positive 3. Since we end up at 7, it follows that $4 + 3 = 7$.

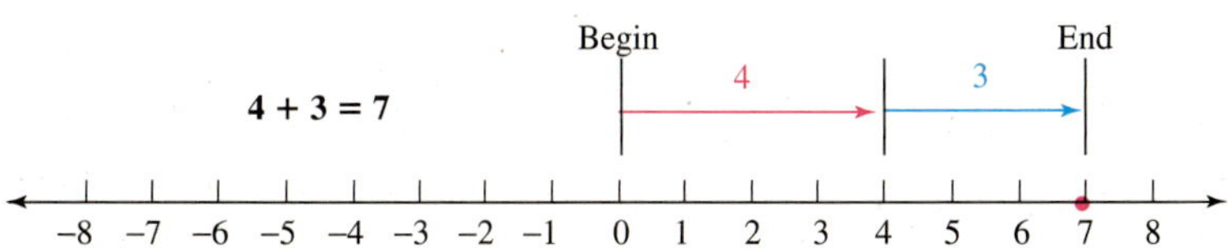

To check our work, let's think of the problem in terms of money. If you won \$4 and won \$3 more, you would have a total of \$7.

To find the sum of two integers with the same sign, both negative, let's consider the problem $-4 + (-3)$. On a number line, we begin at 0 and draw an arrow 4 units long that points to the left. It represents -4. From the tip of that arrow, we draw a second arrow, 3 units long, that points to the left. It represents -3. Since we end up at -7, it follows that $-4 + (-3) = -7$.

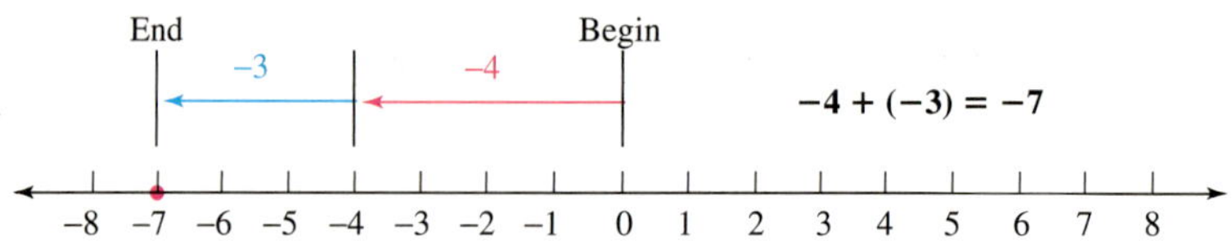

Let's think of this problem in terms of money. If you lost \$4 (-4) and then lost another \$3 (-3), overall, you would have lost a total of \$7 (-7).

Here are some observations about the process of adding two numbers that have the same sign on a number line.

- The arrows representing the integers point in the same direction and they build upon each other.
- The answer has the same sign as the integers that we added.

These observations illustrate the following rules.

Adding Two Integers That Have the Same (Like) Signs

1. To add two positive integers, add them as usual. The final answer is positive.
2. To add two negative integers, add their absolute values and make the final answer negative.

The Language of Mathematics When writing additions that involve integers, write negative integers within parentheses to separate the negative sign $-$ from the plus symbol $+$.

$9 + (-4)$ ~~$9 + -4$~~ and $-9 + (-4)$ ~~$-9 + -4$~~

Self Check 1

Add:

a. $-7 + (-2)$

b. $-25 + (-48)$

c. $-325 + (-169)$

Now Try **Problems 23, 27, and 31**

EXAMPLE 1 Add: **a.** $-3 + (-5)$ **b.** $-26 + (-65)$ **c.** $-456 + (-177)$

Strategy We will use the rule for adding two integers that have the *same sign.*

WHY In each case, we are asked to add two negative integers.

Solution

a. To add two negative integers, we add the absolute values of the integers and make the final answer negative. Since $|-3| = 3$ and $|-5| = 5$, we have

$-3 + (-5) = -8$ Add their absolute values, 3 and 5, to get 8. Then make the final answer negative.

b. Find the absolute values: $|-26| = 26$ and $|-65| = 65$

$-26 + (-65) = -91$ Add their absolute values, 26 and 65, to get 91. Then make the final answer negative.

$$\begin{array}{r} \overset{1}{2}6 \\ +65 \\ \hline 91 \end{array}$$

c. Find the absolute values: $|-456| = 456$ and $|-177| = 177$

$-456 + (-177) = -633$ Add their absolute values, 456 and 177, to get 633. Then make the final answer negative.

$$\begin{array}{r} \overset{1}{4}\overset{1}{5}6 \\ +177 \\ \hline 633 \end{array}$$

Success Tip Calculations that you cannot perform in your head should be shown outside the steps of your solution.

The Language of Mathematics Two negative integers, as well as two positive integers, are said to have *like* signs.

3 Add two integers that have different signs.

Now we will examine the case of adding two integers with different signs. To develop a rule for this situation, let's first consider the problem $4 + (-3)$ on a number line, we begin at 0 and draw an arrow 4 units long that points to the right. This represents positive 4. From the tip of that arrow, we draw a second arrow, 3 units long, that points to the left. It represents -3. Since we end up at 1, it follows that $4 + (-3) = 1$.

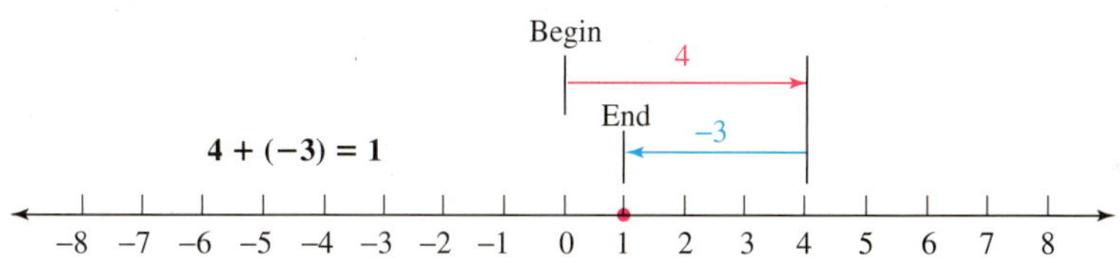

In terms of money, if you won \$4 and then lost \$3 (-3), overall, you would have \$1 left.

Another example will help us better understand how to add two integers with different signs. This time, let's consider the problem $-4 + 3$. On a number line, we begin at 0 and draw an arrow 4 units long that points to the left. It represents -4. From the tip of that arrow, we draw a second arrow, 3 units long, that points to the right. It represents positive 3. Since we end up at -1, it follows that $-4 + 3 = -1$.

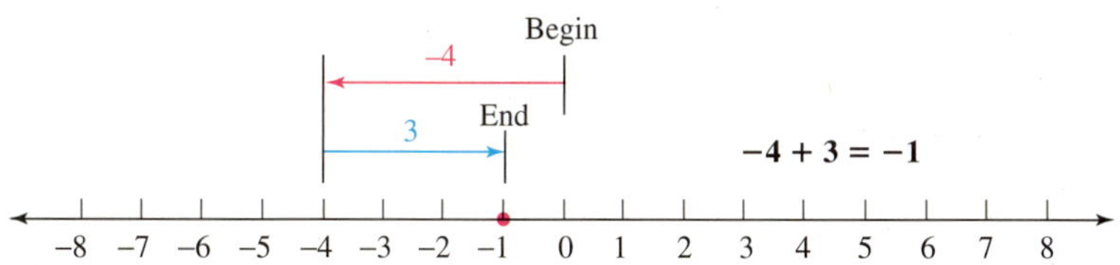

In terms of money, if you lost \$4 (-4) and then won \$3, overall, you have lost \$1 (-1).

Here are some observations about the process of adding two integers that have different signs on a number line.

- The arrows representing the integers point in opposite directions.
- The longer of the two arrows determines the sign of the answer. If the longer arrow represents a positive integer, the sum is positive. If it represents a negative integer, the sum is negative.

These observations suggest the following rules.

Adding Two Integers That Have Different (Unlike) Signs

To add a positive integer and a negative integer, subtract the smaller absolute value from the larger.

1. If the positive integer has the larger absolute value, the final answer is positive.
2. If the negative integer has the larger absolute value, make the final answer negative.

Self Check 2

Add: $6 + (-9)$

Now Try **Problem 35**

EXAMPLE 2 Add: $5 + (-7)$

Strategy We will use the rule for adding two integers that have different signs.

WHY The addend 5 is positive and the addend -7 is negative.

Solution

Step 1 To add two integers with different signs, we first subtract the smaller absolute value from the larger absolute value. Since $|5|$, which is 5, is smaller than $|-7|$, which is 7, we begin by subtracting 5 from 7.

$$7 - 5 = 2$$

Step 2 Since the negative number, -7, has the larger absolute value, we attach a negative sign $-$ to the result from step 1. Therefore,

$$5 + (-7) = -2$$ Make the final answer negative.

The Language of Mathematics A positive integer and a negative integer are said to have *unlike* signs.

Self Check 3

Add:

a. $7 + (-2)$
b. $-53 + 39$
c. $-506 + 888$

Now Try **Problems 37, 39, and 43**

EXAMPLE 3 Add: **a.** $8 + (-4)$ **b.** $-41 + 17$ **c.** $-206 + 568$

Strategy We will use the rule for adding two integers that have different signs.

WHY In each case, we are asked to add a positive integer and a negative integer.

Solution

a. Find the absolute values: $|8| = 8$ and $|-4| = 4$

$8 + (-4) = 4$ Subtract the smaller absolute value from the larger: $8 - 4 = 4$. Since the positive number, 8, has the larger absolute value, the final answer is positive.

b. Find the absolute values: $|-41| = 41$ and $|17| = 17$

$-41 + 17 = -24$ Subtract the smaller absolute value from the larger: $41 - 17 = 24$. Since the negative number, -41, has the larger absolute value, make the final answer negative.

$$\begin{array}{r} \overset{3\,11}{\not{4}\not{1}} \\ -\,17 \\ \hline 24 \end{array}$$

c. Find the absolute values: $|-206| = 206$ and $|568| = 568$

$-206 + 568 = 362$ Subtract the smaller absolute value from the larger: $568 - 206 = 362$. Since the positive number, 568, has the larger absolute value, the answer is positive.

$$\begin{array}{r} 568 \\ -\,206 \\ \hline 362 \end{array}$$

Caution! Did you notice that the answers to the addition problems in Examples 2 and 3 were found using subtraction? This is the case when the addition involves two integers that have *different signs.*

THINK IT THROUGH *Cash Flow*

"College can be trial by fire — a test of how to cope with pressure, freedom, distractions, and a flood of credit card offers. It's easy to get into a cycle of overspending and unnecessary debt as a student."

Planning for College, Wells Fargo Bank

If your income is less than your expenses, you have a *negative* cash flow. A negative cash flow can be a red flag that you should increase your income and/or reduce your expenses. Which of the following activities can increase income and which can decrease expenses?

- Buy generic or store-brand items.
- Get training and/or more education.
- Use your student ID to get discounts at stores, events, etc.
- Work more hours.
- Turn a hobby or skill into a money-making business.
- Tutor young students.
- Stop expensive habits, like smoking, buying snacks every day, etc
- Attend free activities and free or discounted days at local attractions.
- Sell rarely used items, like an old CD player.
- Compare the prices of at least three products or at three stores before buying.

Based on the *Building Financial Skills* by National Endowment for Financial Education.

4 Perform several additions to evaluate expressions.

To **evaluate** (find the value) of expressions that contain several additions, we make repeated use of the rules for adding two integers.

EXAMPLE 4

Evaluate: $-3 + 5 + (-12) + 2$

Strategy Since there are no calculations within parentheses, no exponential expressions, and no multiplication or division, we will perform the additions, working from the left to the right.

WHY This is the next step in the order of operations.

Solution

$-3 + 5 + (-12) + 2 = 2 + (-12) + 2$ Use the rule for adding two integers that have different signs: $-3 + 5 = 2$.

$= -10 + 2$ Use the rule for adding two integers that have different signs: $2 + (-12) = -10$.

$= -8$ Use the rule for adding two integers that have different signs.

Self Check 4

Evaluate: $-12 + 8 + (-6) + 1$

Now Try **Problem 47**

5 Use properties of addition.

The addition of two numbers can be done in any order and the result is the same. For example, $8 + (-1) = 7$ and $-1 + 8 = 7$. This example illustrates that addition is **commutative.**

The Commutative Property of Addition

Changing the order when adding does not affect the answer.
For any real numbers a and b,

$$a + b = b + a$$

The Language of Algebra Commutative is a form of the word *commute,* meaning to go back and forth. *Commuter* trains take people to and from work.

In the following example, we add $-3 + 7 + 5$ in two ways. We will use grouping symbols ( ), called **parentheses,** to show this. Standard practice requires that the operation within the parentheses be performed first.

We read $(-3 + 7) + 5$ as "The quantity of -3 plus 7" pause slightly, and then say "plus 5." We read $-3 + (7 + 5)$ as "-3" pause slightly, and then say "plus the quantity of 7 plus 5." The word **quantity** alerts the reader to the parentheses that are used as grouping symbols.

Method 1: Group −3 and 7

$$(-3 + 7) + 5 = 4 + 5$$
$$= 9$$

Method 2: Group 7 and 5

$$-3 + (7 + 5) = -3 + 12$$
$$= 9$$

It doesn't matter how we group the numbers in this addition; the result is 9. This example illustrates that addition is **associative.**

The Associative Property of Addition

Changing the grouping when adding does not affect the answer.
For any real numbers a, b, and c,

$$(a + b) + c = a + (b + c)$$

Another way to evaluate an expression like that in Example 4 is to use these properties to reorder and regroup the integers in a helpful way. To do the regrouping, we will use grouping symbols called **brackets [ ].** Remember, standard practice requires addition within grouping symbols be performed first.

Self Check 5

Use the commutative and/or associative properties of addition to help evaluate the expression: $-12 + 8 + (-6) + 1$

Now Try **Problem 49**

EXAMPLE 5 Use the commutative and/or associative properties of addition to help evaluate the expression: $-3 + 5 + (-12) + 2$

Strategy We will use the commutative and/or associative properties of addition so that we can add the positives and add the negatives separately. Then we will add those results to obtain the final answer.

WHY It is easier to add integers that have the same sign than integers that have different signs. This approach lessens the possibility of an error, because we only have to add integers that have different signs once.

Solution

$-3 + 5 + (-12) + 2$

$= -3 + (-12) + 5 + 2$ — Use the commutative property of addition to reorder the integers.

Negatives Positives

$= [-3 + (-12)] + (5 + 2)$ — Use the associative property of addition to group the negatives and group the positives.

$= -15 + 7$ Use the rule for adding two integers that have the same sign twice. Add the negatives within the brackets. Add the positives within the parentheses.

$= -8$ Use the rule for adding two integers that have different signs. This is the same result as in Example 4.

Sometimes, an application of the associative property can simplify a calculation.

EXAMPLE 6 Apply the associative property of addition to find the sum:

$-220 + (-80 + 2{,}573)$

Strategy We will use the associative property to group -220 with -80.

WHY It is helpful to regroup because -220 and -80 are a pair of numbers that are easily added.

Solution

$$
\begin{aligned}
-220 + (-80 + 2{,}573) &= (-220 + -80) + 2{,}573 \\
&= -300 + 2{,}573 \quad \text{Do the addition within the parentheses first.} \\
&= 2{,}273
\end{aligned}
$$

$$
\begin{array}{r} 2{,}573 \\ -300 \\ \hline 2{,}273 \end{array}
$$

Self Check 6

Apply the associative property of addition to find the sum: $-140 + (-60 + 4{,}819)$

Now try **Problem 51**

6 Identify opposites (additive inverses) when adding integers.

When 0 is added to a whole number, the whole number remains the same. This is also true for integers. For example, $-5 + 0 = -5$ and $0 + (-43) = -43$. Because of this, we call 0 the **additive identity.**

The Language of Mathematics *Identity* is a form of the word *identical,* meaning the same. You have probably seen *identical* twins.

Addition Property of 0

The sum of any integer and 0 is that integer. For example,

$-3 + 0 = -3, \qquad -19 + 0 = -19, \qquad \text{and} \qquad 0 + (-76) = -76$

There is another important fact about the operation of addition and 0. To illustrate it, we use the number line below to add 6 and its opposite, -6. Notice that $6 + (-6) = 0$.

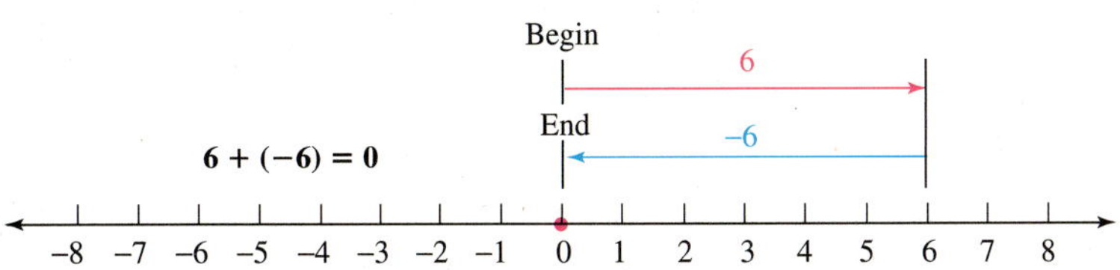

If the sum of two numbers is 0, the numbers are said to be **additive inverses** of each other. Since $6 + (-6) = 0$, we say that 6 and -6 are additive inverses. Likewise, -7 is the additive inverse of 7, and 51 is the additive inverse of -51.

We can now classify a pair of integers such as 6 and -6 in three ways: as opposites, negatives, or additive inverses.

Addition Property of Opposites

The sum of an integer and its opposite (additive inverse) is 0. For example,

$$4 + (-4) = 0, \quad -53 + 53 = 0, \quad \text{and} \quad 710 + (-710) = 0$$

At certain times, the addition property of opposites can be used to make addition of integers easier.

Self Check 7

Evaluate:
$8 + (-1) + 6 + (-8) + 1$

Now Try **Problem 59**

EXAMPLE 7 Evaluate: $12 + (-5) + 6 + 5 + (-12)$

Strategy Instead of working from left to right, we will use the commutative and associative properties of addition to add *pairs of opposites.*

WHY Since the sum of an integer and its opposite is 0, it is helpful to identify such pairs in an addition.

Solution

$$12 + (-5) + 6 + 5 + (-12) = 0 + 0 + 6$$

(12 and −12 are opposites; −5 and 5 are opposites)

Locate pairs of opposites and add them to get 0.

$$= 6$$

The sum of any integer and 0 is that integer.

7 Solve application problems by adding integers.

Since application problems are almost always written in words, the ability to understand what you read is very important. Words and phrases such as *gained, increased by,* and *rise* indicate addition.

Self Check 8

TEMPERATURE CHANGE On the morning of February 21, 1918, in Granville, North Dakota, the morning low temperature was −33°F. By the afternoon, the temperature had risen a record 83 degrees. What was the afternoon high temperature in Granville? (Source: *Extreme Weather* by Christopher C. Burt)

Now Try **Problem 97**

EXAMPLE 8 ***Record Temperature Change*** At the beginning of this section, we learned that at 7:30 A.M. on January 22, 1943, in Spearfish, South Dakota, the temperature was −4°F. The temperature then rose 49 degrees in just 2 minutes. What was the temperature at 7:32 A.M.?

Strategy We will carefully read the problem looking for a key word or phrase.

WHY Key words and phrases indicate what arithmetic operations should be used to solve the problem.

Solution The phrase *rose 49 degrees* indicates addition. With that in mind, we translate the words of the problem, called a **verbal model**, to numbers and symbols.

The temperature at 7:32 A.M.	was	the temperature at 7:30 A.M.	plus	49 degrees.
The temperature at 7:32 A.M.	=	−4	+	49

To find the sum, we will use the rule for adding two integers that have different signs. First, we find the absolute values: $|-4| = 4$ and $|49| = 49$.

$-4 + 49 = 45$ Subtract the smaller absolute value from the larger absolute value: $49 - 4 = 45$. Since the positive number, 49, has the larger absolute value, the final answer is positive.

At 7:32 A.M., the temperature was 45°F.

A number line model can also be helpful to visualize application problems and solve them.

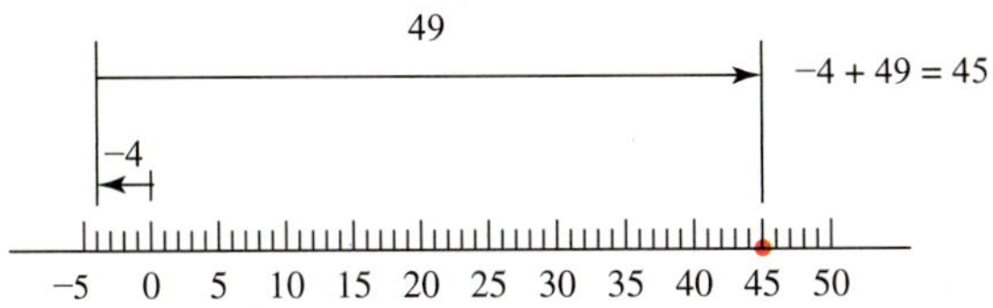

Using Your CALCULATOR Entering Negative Numbers

Canada is the largest U.S. trading partner. To calculate the 2007 U.S. trade balance with Canada, we add the $249 billion worth of U.S. exports *to* Canada (considered positive) to the $317 billion worth of U.S. imports *from* Canada (considered negative). We can use a calculator to perform the addition: 249 + (−317)

We do not have to do anything special to enter a positive number. Negative numbers are entered using either **direct** or **reverse entry,** depending on the type of calculator you have.

To enter −317 using reverse entry, press the change-of-sign key [+/−] *after entering* 317. To enter −317 using direct entry, press the negative key [(−)] *before* entering 317. In either case, note that [+/−] and the [(−)] keys are different from the subtraction key [−].

Reverse entry: 249 [+] 317 [+/−] [=]

Direct entry: 249 [+] [(−)] 317 [ENTER]   −68

In 2007, the United States had a trade balance of −$68 billion with Canada. Because the result is negative, it is called a trade *deficit.*

ANSWERS TO SELF CHECKS

1. a. −9 **b.** −73 **c.** −494 **2.** −3 **3. a.** 5 **b.** −14 **c.** 382 **4.** −9 **5.** −9 **6.** 4,619 **7.** 6 **8.** 50°F

SECTION 1.2 STUDY SET

VOCABULARY

Fill in the blanks.

1. Two negative integers, as well as two positive integers, are said to have the same or ______ signs. A positive integer and a negative integer are said to have different or _______ signs.
2. To evaluate an expression means to find its _______.
3. When 0 is added to a number, the number remains the same. We call 0 the additive ________.
4. Since −5 + 5 = 0, we say that 5 is the additive _______ of −5. We can also say that 5 and −5 are _________.
5. ___________ property of addition: The order in which integers are added does not change their sum.
6. __________ property of addition: The way in which integers are grouped does not change their sum.

CONCEPTS

7. **a.** If you dig a hole 6 feet below ground level, and then fill the hole with 5 feet of dirt, how deep will the hole then be?

   b. Write a number sentence of the form ☐ + ☐ = ☐ to describe the situation in mathematical symbols.

8. **a.** If you lost \$6 and then lost \$8, overall, what amount of money was lost?

b. If you lost \$6 and then won \$8, overall, what amount of money have you won?

9. Illustrate each addition of integers below using a model. Ask your instructor which model (heaps/holes, colored tiles, number line, or money) you should use. Then write a number sentence of the form ☐ + ☐ = ☐ to express the result in mathematical symbols.

a. $5 + (-2)$
b. $-8 + 3$
c. $2 + (-6)$
d. $-4 + (-5)$

10. **a.** On a number line, how far is 11 from 0?

b. On a number line, how far is -11 from 0?

11. **a.** Illustrate the following process using a number line model. Begin at 0 and move 3 units to the right. Then move of 3 units to the left. On what integer does the process end?

b. Illustrate the addition problem described in part a using a heaps/holes, colored tiles, or money model. (Ask your instructor which one you should use.)

c. Write a number sentence of the form ☐ + ☐ = ☐ to express your result to part b in mathematical symbols.

12. **a.** What is the absolute value of 10? What is the absolute value of -12?

b. Which number has the larger absolute value, 10 or -12?

c. Using your answers to part a, subtract the smaller absolute value from the larger absolute value. What is the result?

Fill in the blanks.

13. To add two integers with unlike signs, ________ their absolute values, the smaller from the larger. Then attach to that result the sign of the number with the ______ absolute value.

14. To add two integers with like signs, add their ________ values and attach their common _____ to the sum.

15. **a.** Is the sum of two positive integers always positive?

b. Is the sum of two negative integers always negative?

c. Is the sum of a positive integer and a negative integer always positive?

d. Is the sum of a positive integer and a negative integer always negative?

16. Complete the table by finding the additive inverse, opposite, and absolute value of the given numbers.

Number	Additive inverse	Opposite	Absolute value
19			
−2			
0			

17. **a.** What is the sum of an integer and its additive inverse?

b. What is the sum of an integer and its opposite?

18. **a.** What number must be added to -5 to obtain 0?

b. What number must be added to 8 to obtain 0?

19. Use the commutative property of addition to complete each statement.

a. $-5 + 1 =$ ☐ + ☐
b. $15 + (-80.5) =$ ☐ + ☐
c. $-20 + (4 + 20) = -20 +$ (☐ + ☐)
d. $(2.1 + 3) + 6 =$ (☐ + ☐) $+ 6$

20. Use the associative property of addition to complete each statement.

a. $(-6 + 2) + 8 =$ ☐
b. $-7 + (7 + 3) =$ ☐

NOTATION

Complete each step to evaluate the expression.

21. $-8 + (-2) + 6 =$ ☐ $+ 6$
$=$ ☐

22. $(-3 + 8) + (-3) =$ ☐ $+ (-3)$
$=$ ☐

GUIDED PRACTICE

Add. See Example 1.

23. $-6 + (-3)$
24. $-2 + (-3)$
25. $-5 + (-5)$
26. $-8 + (-8)$
27. $-51 + (-11)$
28. $-43 + (-12)$
29. $-69 + (-27)$
30. $-55 + (-36)$
31. $-248 + (-131)$
32. $-423 + (-164)$
33. $-565 + (-309)$
34. $-709 + (-187)$

Add. See Examples 2 and 3.

35. $-8 + 5$
36. $-9 + 3$
37. $7 + (-6)$
38. $4 + (-2)$
39. $20 + (-42)$
40. $-18 + 10$
41. $71 + (-23)$
42. $75 + (-56)$
43. $479 + (-122)$
44. $589 + (-242)$
45. $-339 + 279$
46. $-704 + 649$

Evaluate each expression. See Examples 4, 5 and 6.

47. $9 + (-3) + 5 + (-4)$

48. $-3 + 7 + (-4) + 1$

49. $6 + (-4) + (-13) + 7$

50. $8 + (-5) + (-10) + 6$

51. $-130 + (-70 + 4{,}911)$

52. $-560 + (-40 + 6{,}823)$

53. $(-355 + 47) + 3$

54. $(-292 + 188) + 2$

55. $-99 + (99 + 215)$

56. $67 + (-67 + 127)$

57. $(-112 + 56) + (-56)$

58. $(-67 + 5) + (-5)$

Evaluate each expression. See Example 7.

59. $23 + (-5) + 3 + 5 + (-23)$

60. $41 + (-1) + 9 + 1 + (-41)$

61. $-10 + (-1) + 10 + (-6) + 1$

62. $-14 + (-30) + 14 + (-9) + 9$

TRY IT YOURSELF

Add.

63. $-2 + 6 + (-1)$

64. $4 + (-3) + (-2)$

65. $-7 + 0$

66. $0 + (-15)$

67. $24 + (-15)$

68. $-4 + 14$

69. $-435 + (-127)$

70. $-346 + (-273)$

71. $-7 + 9$

72. $-3 + 6$

73. $2 + (-2)$

74. $-10 + 10$

75. $2 + (-10 + 8)$

76. $(-9 + 12) + (-4)$

77. $-9 + 1 + (-2) + (-1) + 9$

78. $5 + 4 + (-6) + (-4) + (-5)$

79. $[6 + (-4)] + [8 + (-11)]$

80. $[5 + (-8)] + [9 + (-15)]$

81. $(-4 + 8) + (-11 + 4)$

82. $(-12 + 6) + (-6 + 8)$

83. $-675 + (-456) + 99$

84. $-9{,}750 + (-780) + 2{,}345$

85. Find the sum of -6, -7, and -8.

86. Find the sum of -11, -12, and -13.

87. $-2 + [789 + (-9{,}135)]$

88. $-8 + [2{,}701 + (-4{,}089)]$

89. What is 25 more than -45?

90. What is 31 more than -65?

LOOK ALIKES . . .

91. **a.** $12 + 15$ **b.** $-12 + 15$
c. $-12 + (-15)$ **d.** $12 + (-15)$

92. **a.** $432 + 67$ **b.** $-432 + 67$
c. $-432 + (-67)$ **d.** $432 + (-67)$

CONCEPT EXTENSIONS

In problem 93-96, write a number sentence of the form ☐ + ☐ = ☐ *for each situation.*

93. **a.** The sum of two integers is 12, and one of them is negative.
b. The sum of two integers is -12, and only one of them is negative.
c. The sum of two integers is -12, and both are negative.

94. **a.** The sum of a positive integer and a negative integer is a positive integer.
b. The sum of a positive integer and a negative integer is a negative integer.

95. The sum of two integers is 0.

96. The sum of two negative integers is a negative integer.

APPLICATIONS

Use signed numbers to solve each problem.

97. RECORD TEMPERATURES The lowest recorded temperatures for Michigan and Minnesota are shown below. Use the given information to find the highest recorded temperature for each state.

State	Lowest temperature	Highest temperature
North Carolina	Jan. 21, 1985: −34°F	Aug. 21, 1983: 144 °F warmer than the record low
Minnesota	Feb. 2, 1996: −60°F	July 6, 1936: 174°F warmer than the record low

(Source: *The World Almanac Book of Facts,* 2009)

98. ELEVATIONS The lowest point in the United States is Death Valley, California, with an elevation of -282 feet (282 feet below sea level). Mt. McKinley (Alaska) is the highest point in the United States. Its elevation is 20,602 feet higher than Death Valley. What is the elevation of Mt. McKinley? (Source: *The World Almanac Book of Facts,* 2009)

99. SUNKEN SHIPS Refer to the map below.

a. The German battleship *Bismarck,* one of the most feared warships of World War II, was sunk by the British in 1941. It lies on the ocean floor 15,720 feet below sea level off the west coast of France. Represent that depth using a signed number.

b. In 1912, the famous cruise ship *Titanic* sank after striking an iceberg. It lies on the North Atlantic ocean floor, 3,220 feet higher than the *Bismarck.* At what depth is the *Titanic* resting?

100. JOGGING A businessman's lunchtime workout includes jogging up ten stories of stairs in his high-rise office building. He starts the workout on the fourth level below ground in the underground parking garage.

a. Represent that level using a signed number.

b. On what story of the building will he finish his workout?

101. FLOODING After a heavy rainstorm, a river that had been 9 feet under flood stage rose 11 feet in a 48-hour period.

a. Represent that level of the river before the storm using a signed number.

b. Find the height of the river after the storm in comparison to flood stage.

102. ATOMS An atom is composed of protons, neutrons, and electrons. A proton has a positive charge (represented by $+1$), a neutron has no charge, and an electron has a negative charge (-1). Two simple models of atoms are shown below.

a. How many protons does the atom in figure (a) have? How many electrons?

b. What is the net charge of the atom in figure (a)?

c. How many protons does the atom in figure (b) have? How many electrons?

d. What is the net charge of the atom in figure (b)?

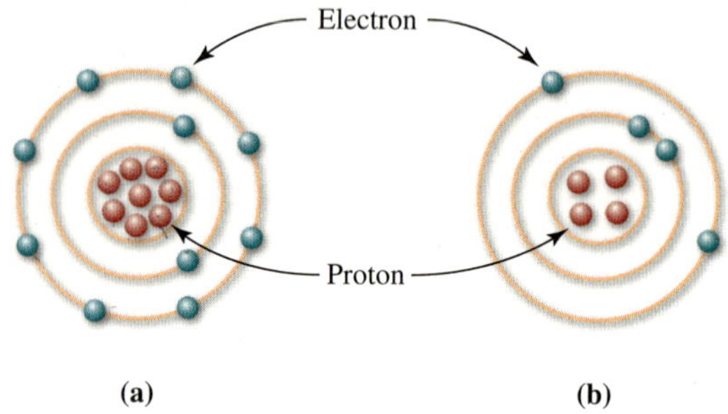

103. CHEMISTRY The three steps of a chemistry lab experiment are listed here. The experiment begins with a compound that is stored at $-40°$F.

Step 1 Raise the temperature of the compound 200°.

Step 2 Add sulfur and then raise the temperature 10°.

Step 3 Add 10 milliliters of water, stir, and raise the temperature 25°.

What is the resulting temperature of the mixture after step 3?

from Campus to Careers
Personal Financial Advisor

104. Suppose as a personal financial advisor, your clients are considering purchasing income property. You find a duplex apartment unit that is for sale and learn that the maintenance costs, utilities, and taxes on it total $900 per month. If the current owner receives monthly rental payments of $450 and $380 from the tenants, does the duplex produce a positive cash flow each month?

105. HEALTH Find the point total for the six risk factors (in blue) on the medical questionnaire. Then use the table at the bottom of the form to determine the patient's risk of contracting heart disease in the next 10 years.

Age		Total Cholesterol	
Age	Points	Reading	Points
35	−4	280	3
Cholesterol		**Blood Pressure**	
HDL	Points	Systolic/Diastolic	Points
62	−3	124/100	3
Diabetic		**Smoker**	
	Points		Points
Yes	4	Yes	2

10-Year Heart Disease Risk

Total Points	Risk	Total Points	Risk
−2 or less	1%	5	4%
−1 to 1	2%	6	6%
2 to 3	3%	7	6%
4	4%	8	7%

Source: National Heart, Lung, and Blood Institute

106. POLITICAL POLLS Six months before a general election, the incumbent senator found himself trailing the challenger by 18 points. To overtake his opponent, the campaign staff decided to use a four-part strategy. Each part of this plan is shown below, with the anticipated point gain.

Part 1 Intense TV ad blitz: gain 10 points

Part 2 Ask for union endorsement: gain 2 points

Part 3 Voter mailing: gain 3 points

Part 4 Get-out-the-vote campaign: gain 1 point

With these gains, will the incumbent overtake the challenger on election day?

107. MILITARY SCIENCE During a battle, an army retreated 1,500 meters, regrouped, and advanced 3,500 meters. The next day, it advanced 1,250 meters. Find the army's net gain.

108. AIRLINES The graph below shows the annual net income for Delta Air Lines during the years 2004–2007.

Find the company's total net income over this span of four years in millions of dollars.

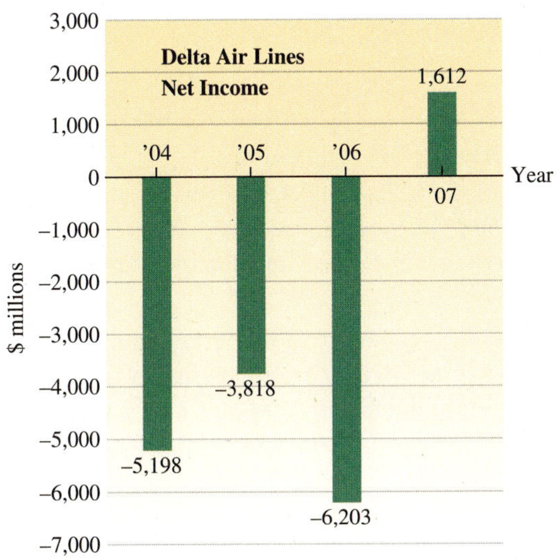

(Source: *The Wall Street Journal*)

109. ACCOUNTING On a financial balance sheet, debts (considered negative numbers) are written within parentheses. Assets (considered positive numbers) are written without parentheses. What is the 2009 fund balance for the preschool whose financial records are shown in the next column?

ABC Preschool Balance Sheet, June 2009

Fund	Balance $
Classroom supplies	$5,889
Emergency needs	$927
Holiday program	($2,928)
Insurance	$1,645
Janitorial	($894)
Licensing	$715
Maintenance	($6,321)
BALANCE	?

110. SPREADSHEETS Monthly rain totals for four counties are listed in the spreadsheet below. The −1 entered in cell B1 means that the rain total for Suffolk County for a certain month was 1 inch *below* average. We can analyze this data by asking the computer to perform various operations.

Book 1

File Edit View Insert Format Tools Data Window Help

	A	B	C	D	E	F
1	Suffolk	−1	−1	0	+1	+1
2	Marin	0	−2	+1	+1	−1
3	Logan	−1	+1	+2	+1	+1
4	Tipton	−2	−2	+1	−1	−3
5						

a. To ask the computer to add the numbers in cells B1, B2, B3, and B4, we type SUM(B1:B4). Find this sum.

b. Find SUM(F1:F4).

111. GAME SHOWS A contestant on *Jeopardy!* Correctly answered the first question to win $100, missed the second to lose $200, correctly answered the third to win $300, and missed the fourth to lose $400. What is her score after answering four questions?

112. COLLEGE MATH CLUBS At the end of the first half of the competition between math clubs, the score for Edgewood College was -100. The club's performance in the second half is shown in the table below. What was their final score in the competition?

Point value of the question	Edgewood's Answer
100	Incorrect
150	Correct
50	Correct
150	Incorrect

113. GOLF The leaderboard below shows the top four finishers from the 2009 PGA Championship Golf Tournament. Scores for each round are compared to *par,* the standard number of strokes necessary to complete the course. A score of -2, for example, indicates that the golfer used two strokes less than par to complete the course. A score of 5 indicates five strokes more than par. Determine the tournament total for each golfer.

Leaderboard

	Round 1	2	3	4	Total
Y.E. Yang	+1	−2	−5	−2	
Tiger Woods	−5	−2	−1	+3	
Lee Westwood	−2	0	+1	−2	
Rory McIlroy	−1	+1	−1	−2	

114. SUBMARINES A submarine was cruising at a depth of 1,250 feet. The captain gave the order to climb 550 feet. Compared to sea level, find the new depth of the sub.

115. MOVIE LOSSES According to the Numbers Box Office Data website, the movie *Stealth,* released in 2005 by Sony Pictures, cost about \$176,350,000 to produce, promote, and distribute. It reportedly earned back just \$76,700,000 worldwide. Express the dollar loss suffered by Sony as a signed number.

116. PHYSICS In the illustration, arrows show the two forces acting on a lamp hanging from a ceiling. What is the sum of the forces?

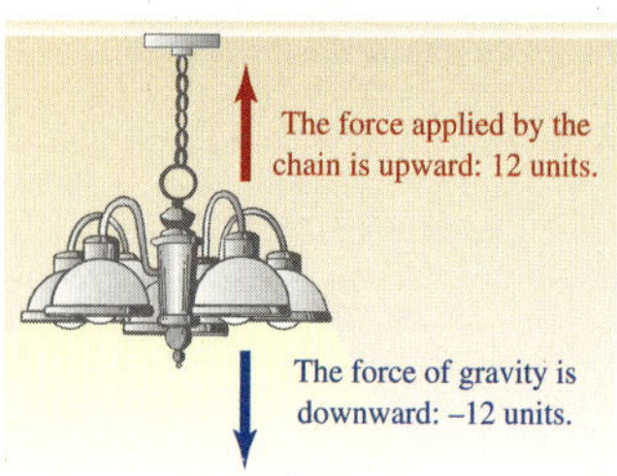

117. ELECTRONICS A closed circuit contains two batteries and three resistors. The sum of the voltages in the loop must be 0. Is it?

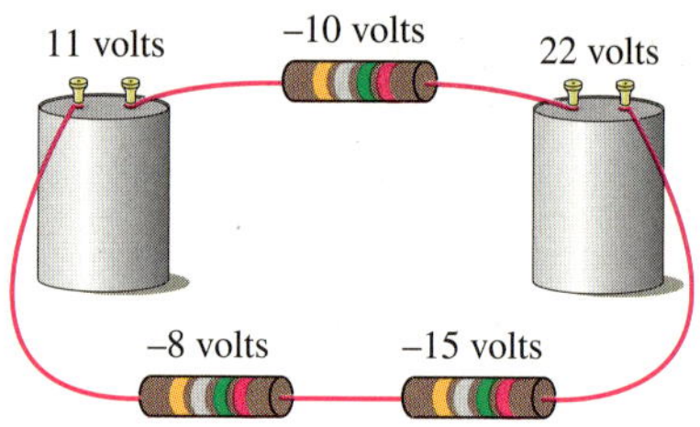

118. THE BIG EASY The city of New Orleans lies, on average, 6 feet below sea level. What is the elevation of the top of an 85-foot tall building in New Orleans?

WRITING

119. Is the sum of a positive and a negative number always positive? Explain why or why not.

120. How do you explain the fact that when asked to *add* -4 and 8, we must actually *subtract* to obtain the result?

121. Explain why the sum of two negative numbers is a negative number.

122. Write an application problem that will require adding -50 and -60.

123. If the sum of two integers is 0, what can be said about the integers? Give an example.

124. Explain why the expression $-6 + -5$ is not written correctly. How should it be written?

125. Explain the commutative property of addition in your own words. Give an example using a positive integer and negative integer.

126. Explain the associative property of addition in your own words. Give an example using two positive integers and a negative integer.

127. What properties were used in Step 1 and Step 2 of the solution?

$$(-99 + 14) + (-1) = [14 + (-99)] + (-1) \quad \text{Step 1}$$
$$= 14 + [-99 + (-1)] \quad \text{Step 2}$$
$$= 14 + (-100)$$
$$= -86$$

SECTION 1.3

Applications Introduction: Subtracting Integers

You have learned in arithmetic that a minus symbol – is used to indicate subtraction. However, in algebra, this same symbol is used in two other ways. Its meaning depends on where it appears in an expression.

$5 - 18$	This is read as "five minus eighteen."
-5	This is usually read as "negative five." It could also be read as "the additive inverse of five" or "the opposite of five."
$-(-5)$	This is usually read as "the opposite of negative five." It could also be read as "the additive inverse of negative five."

To successfully complete Section 1.3, you need to know how to read subtraction problems involving negative numbers correctly.

1. Copy each expression below. Draw a circle ◯ around each negative sign and draw a box ☐ around each minus symbol.

a. $9 - (-3)$ **b.** $-20 - 11$ **c.** $-8 - (-4)$

2. Write each phrase using symbols.

a. One minus negative seven

b. Negative one minus seven

c. Negative one minus negative seven

d. One minus seven

3. GRADING A student's graded math test was returned to her with -15 written in red ink at the top of it. After reviewing the test, she found that her teacher made some mistakes grading it. He marked three answers wrong that were actually correct. When she showed the teacher her test, he admitted his errors, and said he needed to take away three negatives from the -15 he had written on her test. Which expression below could be used find her corrected test score?

$-15 - 3$ $-15 - (-3)$ $-15 + (-3)$ $15 - (-3)$

4. FOREIGN RELATIONS In 2004, Congress forgave \$4 billion of Iraqi debt owed to the United States. Before that, Iraq's total debt was estimated to be \$120 billion. Which expression below represents Iraq's total debt (in billions) after getting debt relief from the U.S.?

$120 + 4$ $120 - (-4)$ $-120 - (-4)$ $120 - 4$

In this section, a horizontal number line is used to explain subtraction of integers. Another way to picture such subtractions is with a **hot air balloon model** and a vertical number line. For example, to model the subtraction $20 - (-1)$, picture a hot air balloon floating at a height of 20 feet above the ground with many sandbags attached to its basket. Since sandbags make the balloon move downward, we will think of each sandbag as representing -1. Imagine *removing* (subtracting) one sandbag from the basket. What would happen? The basket would rise 1 foot, of course, and from the model see that $20 - (-1) = 21$.

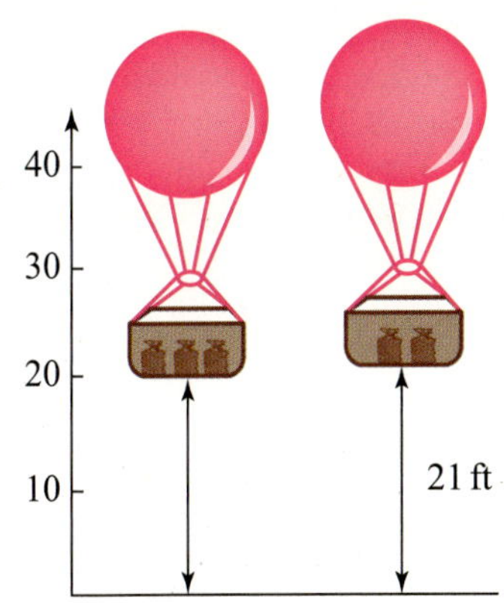

5. Use the hot air balloon model to find each result.

a. $20 - (-2)$ **b.** $100 - (-5)$

c. $250 - (-10)$ **d.** $524 - (-8)$

e. $1{,}500 - (-200)$ **f.** $18{,}000 - (-3{,}000)$

6. Consider the hot air balloon example again. Fill in the blank: We can find the answer to $20 - (-1)$ by _______ positive 1 to 20.

Objectives

1 Use the subtraction rule.

2 Evaluate expressions involving subtraction and addition.

3 Solve application problems by subtracting integers.

SECTION 1.3

Subtracting Integers

ARE YOU READY?

The following problems review some basic concepts that are important when subtracting positive and negative real numbers.

1. What is the opposite of 6? What is the opposite of -15?
2. Write *twenty-two minus six* in symbols.
3. If 8 is subtracted from 20, what is the result?
4. Add: $-11 + 2$

To introduce the concept of subtraction with integers, let's consider the following situation. Suppose you owe a friend \$100. We can represent this debt with the integer -100. Out of kindness, your friend decides to forgive \$20 of that debt. His gift, in effect is taking away (subtracting) -20 from the dollar amount you originally owed him. We can describe the forgiveness of debt in symbols using subtraction:

$$-100 - (-20)$$

However, forgiving (or taking away) a debt has the same effect as giving someone money. Therefore, we could also describe this situation in symbols using addition:

$$-100 + 20$$

In this section, we will use this relationship between subtraction and addition to develop a rule that is very helpful when subtracting integers.

1 Use the subtraction rule.

The subtraction problem $6 - 4$ can be thought of as taking away 4 from 6. We can use a number line to illustrate this. Beginning at 0, we draw an arrow of length 6 units long that points to the right. It represents positive 6. From the tip of that arrow, we draw a second arrow, 4 units long, that points to the left. It represents taking away 4. Since we end up at 2, it follows that $6 - 4 = 2$.

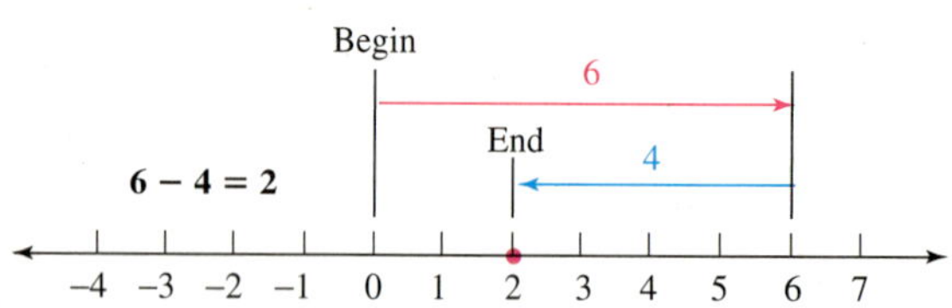

Note that the illustration above also represents the *addition* $6 + (-4) = 2$. We see that

Subtracting 4 from 6 . . . is the same as . . . adding the opposite of 4 to 6.

$$6 - 4 = 2 \qquad\qquad 6 + (-4) = 2$$

The results are the same.

This observation suggests the following rule.

Rule for Subtraction

To subtract two integers, add the first integer to the opposite (additive inverse) of the integer to be subtracted.

Put more simply, this rule says that **subtraction is the same as adding the opposite.**

After rewriting a subtraction as addition of the opposite, we then use one of the rules for the addition of signed numbers to find the result.

You won't need to use this rule for every subtraction problem. For example, $6 - 4$ is obviously 2; it does not need to be rewritten as adding the opposite. But for more complicated problems such as $-6 - 4$ or $3 - (-5)$, where the result is not obvious, the subtraction rule will be quite helpful.

EXAMPLE 1 Subtract and check the result:

a. $-6 - 4$ **b.** $3 - (-5)$ **c.** $7 - 23$

Strategy To find each difference, we will apply the rule for subtraction: Add the first integer to the opposite of the integer to be subtracted.

WHY It is easy to make an error when subtracting signed numbers. We will probably be more accurate if we write each subtraction as addition of the opposite.

Solution

a. We read $-6 - 4$ as "negative six *minus* four." Thus, the number to be subtracted is 4. Subtracting 4 is the same as adding its opposite, -4.

Change the subtraction to addition.

$$-6 - 4 = -6 + (-4) = -10$$

Use the rule for adding two integers with the same sign.

Change the number being subtracted to its opposite.

To check, we add the *difference,* -10, and the *subtrahend,* 4. We should get the *minuend,* -6.

Check: $-10 + 4 = -6$ The result checks.

Caution! Don't forget to write the opposite of the number to be subtracted within parentheses if it is negative.

$$-6 - 4 = -6 + (-4)$$

b. We read $3 - (-5)$ as "three *minus* negative five." Thus, the number to be subtracted is -5. Subtracting -5 is the same as adding its opposite, 5.

Add . . .

$$3 - (-5) = 3 + 5 = 8$$

. . . the opposite

Check: $8 + (-5) = 3$ The result checks.

Self Check 1

Subtract and check the result:

a. $-2 - 3$

b. $4 - (-8)$

c. $6 - 85$

Now Try **Problems 21, 25, and 29**

c. We read $7 - 23$ as "seven *minus* twenty-three." Thus, the number to be subtracted is 23. Subtracting 23 is the same as adding its opposite, -23.

Add . . .

$$7 - 23 = 7 + (-23) = -16$$ Use the rule for adding two integers with different signs.

. . . the opposite

Check: $-16 + 23 = 7$ The result checks.

Caution! When applying the subtraction rule, *do not change* the first number.

$$-6 - 4 = -6 + (-4) \qquad 3 - (-5) = 3 + 5$$

Self Check 2

a. Subtract -10 from -7.

b. Subtract -7 from -10.

Now Try **Problem 33**

EXAMPLE 2 **a.** Subtract -12 from -8. **b.** Subtract -8 from -12.

Strategy We will translate each phrase to mathematical symbols and then perform the subtraction. We must be careful when translating the instruction to subtract one number *from* another number.

WHY The order of the numbers in each word phrase must be reversed when we translate it to mathematical symbols.

Solution

a. Since -12 is the number to be subtracted, we reverse the order in which -12 and -8 appear in the sentence when translating to symbols.

Subtract -12 from -8

$$-8 - (-12)$$ Write -12 within parentheses.

To find this difference, we write the subtraction as addition of the opposite:

Add . . .

$$-8 - (-12) = -8 + 12 = 4$$ Use the rule for adding two integers with different signs.

. . . the opposite

b. Since -8 is the number to be subtracted, we reverse the order in which -8 and -12 appear in the sentence when translating to symbols.

Subtract -8 from -12

$$-12 - (-8)$$ Write -8 within parentheses.

To find this difference, we write the subtraction as addition of the opposite:

Add . . .

$$-12 - (-8) = -12 + 8 = -4$$ Use the rule for adding two integers with different signs.

. . . the opposite

Caution! When subtracting two numbers it is important that we write them in the correct order, because subtraction is not commutative. For instance, in Example 2 parts a and b, note that $-8 - (-12) \neq -12 - (-8)$.

Remember that any subtraction problem can be rewritten as an equivalent addition. We just add the opposite of the number that is to be subtracted. Here are four examples:

- $4 - 8 = 4 + (-8) = -4$
- $4 - (-8) = 4 + 8 = 12$
- $-4 - 8 = -4 + (-8) = -12$
- $-4 - (-8) = -4 + 8 = 4$

Any subtraction can be written as addition of the opposite of the number to be subtracted.

> ***The Language of Mathematics*** When we change a number to its opposite, we say we have *changed* (or *reversed*) its sign.

2 Evaluate expressions involving subtraction and addition.

Expressions can involve repeated subtraction or combinations of subtraction and addition. To **evaluate** (find the value of) them, we use the order of operations rule.

EXAMPLE 3

Evaluate: $-1 - (-2) - 10$

Strategy This expression involves two subtractions. We will write each subtraction as addition of the opposite and then evaluate the expression using the order of operations rule.

WHY It is easy to make an error when subtracting signed numbers. We will probably be more accurate if we write each subtraction as addition of the opposite.

Solution We apply the rule for subtraction twice and then perform the additions, working from left to right. (We could also add the positives and the negatives separately, and then add those results.)

$-1 - (-2) - 10 = -1 + 2 + (-10)$ Add the opposite of −2, which is 2. Add the opposite of 10, which is −10.

$= 1 + (-10)$ Work from left to right. Add −1 + 2 using the rule for adding integers that have different signs.

$= -9$ Use the rule for adding integers that have different signs.

Self Check 3

Evaluate: $-3 - 5 - (-1)$

Now Try **Problem 37**

EXAMPLE 4

Evaluate: $-80 - (-2 - 24)$

Strategy We will consider the subtraction within the parentheses first and rewrite it as addition of the opposite.

WHY By the order of operations rule, we must perform all calculations within parentheses first.

Solution

$-80 - (-2 - 24) = -80 - [-2 + (-24)]$ Add the opposite of 24, which is −24. Since −24 must be written within parentheses, we write −2 + (−24) within brackets.

$= -80 - (-26)$ Within the brackets, add −2 and −24. Since only one set of grouping symbols is now needed, we can write the answer, −26, within parentheses.

$= -80 + 26$ Add the opposite of −26, which is 26.

$= -54$ Use the rule for adding integers that have different signs.

$$\begin{array}{r} \overset{7\,10}{\cancel{8}\cancel{0}} \\ -\ 26 \\ \hline 54 \end{array}$$

Self Check 4

Evaluate: $-72 - (-6 - 51)$

Now Try **Problem 49**

Self Check 5

Evaluate:
$-(-3) + (-16) - 9 - (-28)$

Now Try **Problem 55**

EXAMPLE 5 Evaluate: $-(-6) + (-18) - 4 - (-51)$

Strategy This expression involves one addition and two subtractions. We will write each subtraction as addition of the opposite and then evaluate the expression.

WHY It is easy to make an error when subtracting signed numbers. We will probably be more accurate if we write each subtraction as addition of the opposite.

Solution We apply the rule for subtraction twice. Then we will add the positives and the negatives separately, and add those results. (By the commutative and associative properties of addition, we can add the integers in any order.)

$-(-6) + (-18) - 4 - (-51)$

$= 6 + (-18) + (-4) + 51$ Simplify: $-(-6) = 6$. Add the opposite of 4, which is -4, and add the opposite of -51, which is 51.

$= (6 + 51) + [(-18) + (-4)]$ Reorder the integers. Then group the positives together and group the negatives together.

$= 57 + (-22)$ Add the positives within the parentheses. Add the negatives within the brackets.

$= 35$ Use the rule for adding integers that have different signs.

3 Solve application problems by subtracting integers.

Subtraction finds the *difference* between two numbers. When we find the difference between the maximum value and the minimum value of a collection of measurements, we are finding the **range** of the values.

Range = maximum value − minimum value

Self Check 6

THE GATEWAY CITY The record high temperature for St. Louis, Missouri, is 107°F. The record low temperature is −18°F. Find the temperature range for these extremes. (Source: *The World Almanac and Book of Facts,* 2009)

Now Try **Problem 107**

EXAMPLE 6 ***The Windy City*** The record high temperature for Chicago, Illinois, is 104°F. The record low is −27°F. Find the temperature range for these extremes. (Source: *The World Almanac and Book of Facts,* 2009)

Strategy We will subtract the lowest temperature (−27°F) from the highest temperature (104°F).

WHY The *range* of a collection of data indicates the spread of the data. It is the difference between the largest and smallest values.

Solution We apply the rule for subtraction and add the opposite of −27.

$104 - (-27) = 104 + 27$ 104° is the highest temperature and −27° is the lowest.

$= 131$

The temperature range for these extremes is 131°F.

If we represent the temperatures on a horizontal number line, we see that the temperature range is 27 units + 104 units = 131 units. This verifies the result from the calculations above.

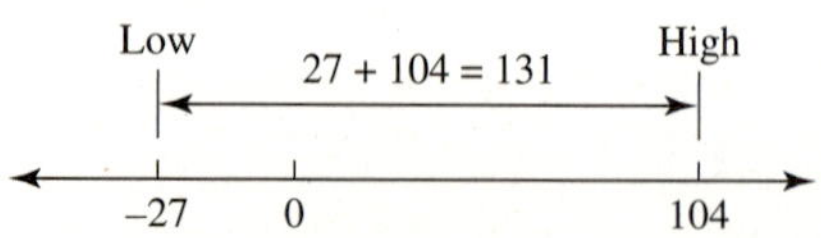

Things are constantly changing in our daily lives. The amount of money we have in the bank, the price of gasoline, and our ages are examples. In mathematics, the operation of subtraction is used to measure change. To find the **change** in a quantity, we subtract the earlier value from the later value.

Change = later value − earlier value

EXAMPLE 7 ***Water Management***

Lake Mead, on the Nevada-Arizona border, is the largest reservoir in the United States. It is formed by Hoover Dam across the Colorado River. In 2000, the water level in Lake Mead was 89 feet above drought level. By 2010, the water level was 27 feet below drought level. Find the change in the water level over that time span. (Source: Bureau of Reclamation)

©iStockphoto.com/MvH

Strategy We can represent a water level above drought level using a positive number and a water level below drought level using a negative number. To find the change in the water level, we will subtract.

WHY In general, *to find the change in a quantity, we subtract the earlier value from the later value.*

Solution

$-27 - 89 = -27 + (-89)$ The earlier water level in 2000 (89 ft) is subtracted from the later water level in 2010 (−27 ft).

$= -116$ Do the addition.

The negative result indicates that the water level *fell* 116 feet in that time span.

Self Check 7

WATER MANAGEMENT Find the change in water level for a week that started at 4 feet above normal and went to 7 feet below normal level.

Now Try **Problem 111**

Using Your CALCULATOR **Subtraction with Negative Numbers**

The world's highest peak is Mount Everest in the Himalayas. The greatest ocean depth yet measured lies in the Mariana Trench near the island of Guam in the western Pacific. To find the range between the highest peak and the greatest depth, we must subtract:

$29{,}035 - (-36{,}025)$

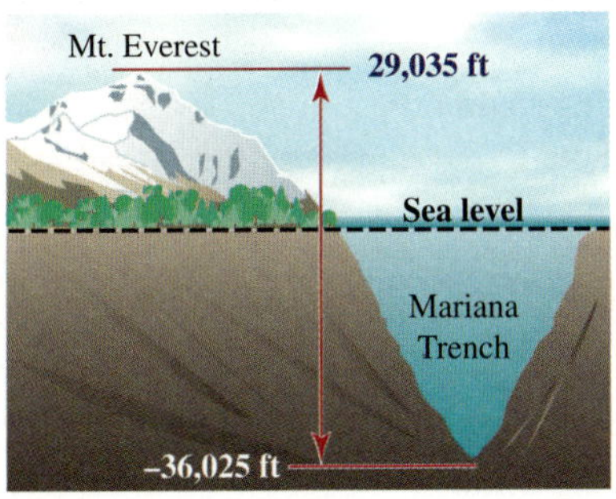

To perform this subtraction on a calculator, we enter the following:

Reverse entry: 29035 [−] 36025 [+/−] [=]

Direct entry: 29035 [−] [(−)] 36025 [ENTER] [65060]

The range is 65,060 feet between the highest peak and the lowest depth. (We could also write 29,035 − (−36,025) as 29,035 + 36,025 and then use the addition key [+] to find the answer.)

ANSWERS TO SELF CHECKS

1. a. −5 **b.** 12 **c.** −79 **2. a.** 3 **b.** −3 **3.** −7 **4.** −15 **5.** 6 **6.** 125°F
7. The water level fell 11 ft.

SECTION 1.3 STUDY SET

VOCABULARY

Fill in the blanks.

1. -8 is the ________ (or ________ inverse) of 8.
2. When we change a number to its opposite, we say we have *changed* (or *reversed*) its _____.
3. To evaluate an expression means to find its ______.
4. The difference between the maximum and the minimum value of a collection of measurements is called the ______ of the values.

CONCEPTS

Fill in the blanks.

5. To subtract two integers, add the first integer to the ________ (additive inverse) of the integer to be subtracted.
6. Subtracting is the same as _______ the opposite.
7. Subtracting 3 is the same as adding $\square$.
8. Subtracting -6 is the same as adding $\square$.
9. We can find the _______ in a quantity by subtracting the earlier value from the later value.
10. After rewriting a subtraction as addition of the opposite, we then use one of the rules for the ________ of signed numbers to find the result.
11. In each case, determine what number is being subtracted.
    a. $-7 - 3$ b. $1 - (-12)$
12. Fill in the blanks to rewrite each subtraction as addition of the opposite of the number being subtracted.
    a. $2 - 7 = 2 + \square$
    b. $2 - (-7) = 2 + \square$
    c. $-2 - 7 = -2 + \square$
    d. $-2 - (-7) = -2 + \square$
13. Apply the rule for subtraction and fill in the three blanks.

    $3 - (-6) = 3 \;\square\; \square = \square$

14. Use addition to check this subtraction: $14 - (-2) = 12$. Is the result correct?

NOTATION

15. Write each phrase using symbols.
    a. negative eight minus negative four
    b. negative eight subtracted from negative four
16. Write each phrase in words.
    a. $7 - (-2)$
    b. $-2 - (-7)$

Complete each solution to evaluate each expression.

17. $1 - 3 - (-2) = 1 + (\square) + 2$
    $= -2 + \square$
    $= \square$
18. $-6 + 5 - (-5) = -6 + 5 + \square$
    $= \square + 5$
    $= \square$
19. $(-8 - 2) - (-6) = [-8 + (\square)] - (-6)$
    $= \square - (-6)$
    $= -10 + \square$
    $= \square$
20. $-(-5) - (-1 - 4) = \square - [-1 + (\square)]$
    $= 5 - (\square)$
    $= 5 + \square$
    $= \square$

GUIDED PRACTICE

Subtract. See Example 1.

21. $-4 - 3$
22. $-4 - 1$
23. $-5 - 5$
24. $-7 - 7$
25. $8 - (-1)$
26. $3 - (-8)$
27. $11 - (-7)$
28. $10 - (-5)$
29. $3 - 21$
30. $8 - 32$
31. $15 - 65$
32. $12 - 82$

Perform the indicated operation. See Example 2.

33. a. Subtract -1 from -11.
    b. Subtract -11 from -1.
34. a. Subtract -2 from -19.
    b. Subtract -19 from -2.
35. a. Subtract -41 from -16.
    b. Subtract -16 from -41.
36. a. Subtract -57 from -15.
    b. Subtract -15 from -57.

Evaluate each expression. See Example 3.

37. $-4 - (-4) - 15$
38. $-3 - (-3) - 10$
39. $10 - 9 - (-8)$
40. $16 - 14 - (-9)$

41. $-1 - (-3) - 4$

42. $-2 - 4 - (-1)$

43. $-5 - 8 - (-3)$

44. $-6 - 5 - (-1)$

Evaluate each expression. See Example 4.

45. $-1 - (-4 - 6)$

46. $-7 - (-2 - 14)$

47. $-42 - (-16 - 14)$

48. $-45 - (-8 - 32)$

49. $-9 - (6 - 7)$

50. $-13 - (6 - 12)$

51. $-8 - (4 - 12)$

52. $-9 - (1 - 10)$

Evaluate each expression. See Example 5.

53. $-(-5) + (-15) - 6 - (-48)$

54. $-(-2) + (-30) - 3 - (-66)$

55. $-(-3) + (-41) - 7 - (-19)$

56. $-(-1) + (-52) - 4 - (-21)$

Use a calculator to perform each subtraction. See *Using Your Calculator.*

57. $-1{,}557 - 890$

58. $20{,}007 - (-496)$

59. $-979 - (-44{,}879)$

60. $-787 - 1{,}654 - (-232)$

TRY IT YOURSELF

Evaluate each expression.

61. $5 - 9 - (-7)$

62. $6 - 8 - (-4)$

63. Subtract -3 from 7.

64. Subtract 8 from -2.

65. $-2 - (-10)$

66. $-6 - (-12)$

67. $0 - (-5)$

68. $0 - 8$

69. $(6 - 4) - (1 - 2)$

70. $(5 - 3) - (4 - 6)$

71. $-5 - (-4)$

72. $-9 - (-1)$

73. $-3 - 3 - 3$

74. $-1 - 1 - 1$

75. $-(-9) + (-20) - 14 - (-3)$

76. $-(-8) + (-33) - 7 - (-21)$

77. $[-4 + (-8)] - (-6) + 15$

78. $[-5 + (-4)] - (-2) + 22$

79. Subtract -6 from -10.

80. Subtract -4 from -9.

81. $-3 - (-3)$

82. $-5 - (-5)$

83. $-8 - [4 - (-6)]$

84. $-1 - [5 - (-2)]$

85. $4 - (-4)$

86. $-3 - 3$

87. $(-6 - 5) - 3 + (-11)$

88. $(-2 - 1) - 5 + (-19)$

LOOK ALIKES . . .

89. **a.** $-50 + (-3)$ **b.** $-50 - (-3)$

90. **a.** $296 + (-178)$ **b.** $296 - (-178)$

CONCEPT EXTENSIONS

In problems 91 and 92, write a number sentence of the form □ − □ = □ *for each situation.*

91. **a.** The difference of two positive integers is -3.

b. The difference of two negative integers is -3.

92. **a.** The difference of two positive integers is 10.

b. The difference of two negative integers is -10.

93. Find the values for points A, B, and C on the number line below. Then write a number sentence that calculates each change.

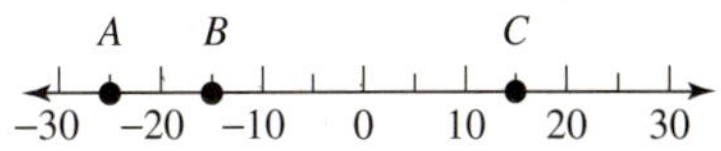

a. From A to B

b. From C to A

94. Suppose x is positive and y is negative. Determine whether each statement is true or false.

a. $x - y > 0$

b. $y - x < 0$

c. $|-x| < 0$

d. $-|y| < 0$

APPLICATIONS

Use signed numbers to solve each problem.

95. SUBMARINES A submarine was traveling 2,000 feet below the ocean's surface when the radar system warned of a possible collision with another sub. The captain ordered the navigator to dive an additional 200 feet and then level off. Find the depth of the submarine after the dive.

96. SCUBA DIVING A diver jumps from his boat into the water and descends to a depth of 50 feet. He pauses to check his equipment and then descends an additional 70 feet. Use a signed number to represent the diver's final depth.

97. GEOGRAPHY The elevation of Death Valley, California, is 282 feet below sea level. The elevation of the Dead Sea in Israel is 1,312 feet below sea level. Find the difference in their elevations.

98. HISTORY Two of the greatest Greek mathematicians were Archimedes (287–212 B.C.) and Pythagoras (569–500 B.C.).

a. Express the year of Archimedes' birth as a negative number.

b. Express the year of Pythagoras' birth as a negative number.

c. How many years apart were they born?

99. AMPERAGE During normal operation, the ammeter on a car reads +5. If the headlights are

turned on, they lower the ammeter reading 7 amps. If the radio is turned on, it lowers the reading 6 amps. What number will the ammeter register if they are both turned on?

100. GIN RUMMY After a losing round, a card player must deduct the value of each of the cards left in his hand from his previous point total of 21. If face cards are counted as 10 points, what is his new score?

101. FOOTBALL A college football team records the outcome of each of its plays during a game on a stat sheet. Find the net gain (or loss) after the third play.

Down	Play	Result
1st	Run	Lost 1 yd
2nd	Pass—sack!	Lost 6 yd
Penalty	Delay of game	Lost 5 yd
3rd	Pass	Gained 8 yd

102. ACCOUNTING Complete the balance sheet below. Then determine the overall financial condition of the company by subtracting the total debts from the total assets.

Walker Corporation
Balance Sheet 2010

Assets	
Cash	$11 1 0 9
Supplies	7 8 6 2
Land	67 5 4 3
Total assets	$
Debts	
Accounts payable	$79 0 3 7
Income taxes	20 1 8 1
Total debts	$

103. OVERDRAFT PROTECTION A student forgot that she had only $15 in her bank account and wrote a check for $25, used an ATM to get $40 cash, and used her debit card to buy $30 worth of groceries. On each of the three transactions, the bank charged her a $20 overdraft protection fee. Find the new account balance.

104. CHECKING ACCOUNTS Michael has $1,303 in his checking account. Can he pay his car insurance premium of $676, his utility bills of $121, and his rent of $750 without having to make another deposit? Explain.

105. TEMPERATURE EXTREMES The highest and lowest temperatures ever recorded in several cities are shown below. List the cities in order, from the largest to smallest range in temperature extremes.

Extreme Temperatures

City	Highest	Lowest
Atlantic City, NJ	102	−3
Barrow, AK	79	−56
Greensboro, NC	104	−8
Norfolk, VA	105	−3
Portland, ME	103	−39

106. EYESIGHT *Nearsightedness,* the condition where near objects are clear and far objects are blurry, is measured using negative numbers. Farsightedness, the condition where far objects are clear and near objects are blurry, is measured using positive numbers. Find the range in the measurements shown in the next column.

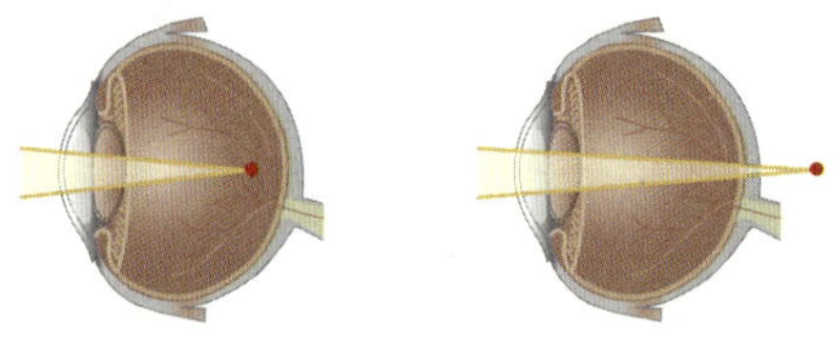

Nearsighted −2 Farsighted +4

107. FREEZE DRYING To make freeze-dried coffee, the coffee beans are roasted at a temperature of 360°F and then the ground coffee bean mixture is frozen at a temperature of −110°F. What is the temperature range of the freeze-drying process?

© Tony Freeman/Photo Edit

108. WEATHER Rashawn flew from his New York home to Hawaii for a week of vacation. He left blizzard conditions and a temperature of −6°F, and stepped off the airplane into 85°F weather. What temperature change did he experience?

109. READING PROGRAMS In a state reading test given at the start of a school year, an elementary school's performance was 23 points below the county average. The principal immediately began a special tutorial program. At the end of the school year, retesting showed the students to be only 7 points below the average. How did the school's reading score change over the year?

110. LIE DETECTOR TESTS On one lie detector test, a burglar scored −18, which indicates deception. However, on a second test, he scored −1, which is inconclusive. Find the change in his scores.

111. U.S. JOBS The table lists the three occupations that are predicted to have the largest job declines from 2008 to 2018. Complete the column labeled "Change."

Number of jobs

Occupation	2008	2018	Change
Farmers/ranchers	985,900	906,700	
Sewing machine operators	212,400	140,900	
Order clerks	245,700	181,500	

Source: Bureau of Labor Statistics

112. GAUGES With the engine off, the ammeter on a car reads 0. If the headlights, which draw a current of 7 amps, and the radio, which draws a current of 6 amps, are both turned on, what will be the new reading?

113. CARD GAMES Gonzalo won the second round of a card game and earned 50 points. Matt and Hydecki had to deduct the value of each of the cards left in their hands from their score on the first round. Use this information to update the score sheet below. (Face cards are counted as 10 points, aces as 1 point, and all others have the value of the number printed on the card.)

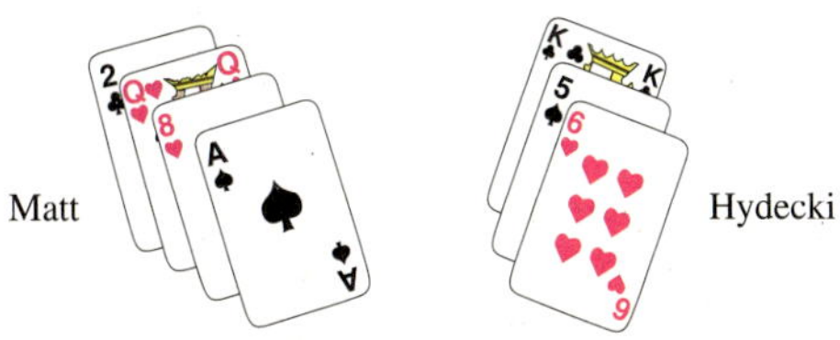

Running point total	Round 1	Round 2
Matt	+50	
Gonzalo	−15	
Hydecki	−2	

114. WORLD'S COLDEST ICE CREAM Dippin' Dots is an ice cream snack that was invented by Curt Jones in 1987. The tiny multi-colored beads are created by flash freezing ice cream mix in liquid nitrogen at a temperature of −355°F. When they come out of the processor, they are stored at a temperature of −40°F. Find the change in temperature of Dippin' Dots from production to storage. (Source: fundinguniverse.com)

115. HISTORY Plato, a famous Greek philosopher, died in 347 B.C. at the age of 81. When was he born?

116. PHYSICS The illustration shows an example of a *standing wave.* What is the difference in the height of the crest of the wave and the depth of the trough of the wave?

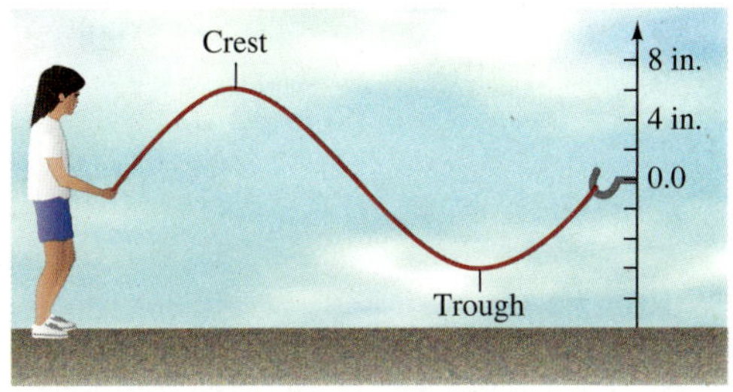

WRITING

117. Explain what is meant when we say that subtraction is the same as addition of the opposite.

118. Give an example showing that it is possible to subtract something from nothing.

119. Explain how to check the result: $-7 - 4 = -11$

120. Explain why students don't need to change every subtraction they encounter to an addition of the opposite. Give some examples.

121. Why is addition of signed numbers taught before subtraction of signed numbers?

122. Explain why we know that the answer to $4 - 10$ is negative without having to do any computation.

123. Is the following statement true or false? Explain. *Having a debt of $100 forgiven is equivalent to gaining $100.*

124. Explain why the operation of subtraction is not commutative.

125. Explain how addition can be used to check subtraction.

Objectives

1. Multiply two integers that have different signs.
2. Multiply two integers that have the same sign.
3. Perform several multiplications to evaluate expressions.
4. Solve application problems by multiplying integers.

SECTION 1.4

Multiplying Integers

ARE YOU READY?

The following problems review some basic concepts that are important when multiplying integers.

1. Find $|-14|$ and $|67|$
2. Do the integers -3 and 24 have the same sign or different signs?
3. Add: $-3 + (-3) + (-3) + (-3)$
4. $\begin{array}{r} 846 \\ \times\ 79 \\ \hline \end{array}$
5. Multiply: 323(100)
6. Multiply: 5(3)(4)

Multiplication of integers is very much like multiplication of whole numbers. The only difference is that we must determine whether the answer is positive or negative.

When we multiply two nonzero integers, they either have different signs or they have the same sign. This means that there are two possibilities to consider.

1 Multiply two integers that have different signs.

To develop a rule for multiplying two integers that have different signs, we will find $4(-3)$, which is the product of a positive integer and negative integer. We say that the signs of the factors are *unlike*. Recall that multiplication is *repeated addition*. Therefore, $4(-3)$ means that we are to add -3 four times.

$$4(-3) = (-3) + (-3) + (-3) + (-3)$$ Write -3 as an addend four times.

$$= -12$$ Use the rule for adding integers that have the same sign.

The result is negative. As a check, think in terms of money. If you lose \$3 four times, you have lost a total of \$12, which is written $-\$12$. This example illustrates the following rule.

Multiplying Two Integers That Have Different (Unlike) Signs

To multiply a positive integer and a negative integer, multiply their absolute values. Then make the final answer negative.

The Language of Algebra The names of the parts of a multiplication fact are:

Factor Factor Product

$$4(-3) = 12$$

Self Check 1

Multiply:

a. $2(-6)$

b. $30(-4)$

c. $-75 \cdot 17$

d. $-98(1{,}000)$

Now Try **Problems 21, 25, 29, and 31**

EXAMPLE 1 Multiply:

a. $7(-5)$ **b.** $20(-8)$ **c.** $-93 \cdot 16$ **d.** $-34(1{,}000)$

Strategy We will use the rule for multiplying two integers that have different (unlike) signs.

WHY In each case, we are asked to multiply a positive integer and a negative integer.

Solution

a. Find the absolute values: $|7| = 7$ and $|-5| = 5$.

$$7(-5) = -35$$ Multiply the absolute values, 7 and 5, to get 35. Then make the final answer negative.

b. Find the absolute values: $|20| = 20$ and $|-8| = 8$.

$20(-8) = -160$ Multiply the absolute values, 20 and 8, to get 160. Then make the final answer negative.

c. Find the absolute values: $|-93| = 93$ and $|16| = 16$.

$-93 \cdot 16 = -1{,}488$ Multiply the absolute values, 93 and 16, to get 1,488. Then make the final answer negative.

$$\begin{array}{r} 93 \\ \times\ 16 \\ \hline 558 \\ 930 \\ \hline 1{,}488 \end{array}$$

d. To find the product of a whole number and 10, 100, 1,000, and so on, *attach the number of zeros in that number to the right of the whole number.* This rule can be extended to products of integers and 10, 100, 1,000, and so on.

$-34(1{,}000) = -34{,}000$ Since 1,000 has three zeros, attach three 0's after −34.

Caution! When writing multiplication involving signed numbers, do not write a negative sign − next to a raised dot · (the multiplication symbol). Instead, use parentheses to show the multiplication.

$6(-2)$ ~~$6 \cdot -2$~~ and $-6(-2)$ ~~$-6 \cdot -2$~~

2 Multiply two integers that have the same sign.

To develop a rule for multiplying two integers that have the same sign, we will first consider 4(3), which is the product of two positive integers. We say that the signs of the factors are *like.* By the definition of multiplication, 4(3) means that we are to add 3 four times.

$4(3) = 3 + 3 + 3 + 3$ Write 3 as an addend four times.

$= 12$ The result is 12, which is a positive number.

As expected, the result is positive.

To develop a rule for multiplying two negative integers, consider the following list, where we multiply −4 by factors that decrease by 1. We know how to find the first four products. Graphing those results on a number line is helpful in determining the last three products.

This factor decreases by 1 each time. Look for a pattern here.

$$\begin{aligned} -4(3) &= -12 \\ -4(2) &= -8 \\ -4(1) &= -4 \\ -4(0) &= 0 \\ -4(-1) &= ? \\ -4(-2) &= ? \\ -4(-3) &= ? \end{aligned}$$

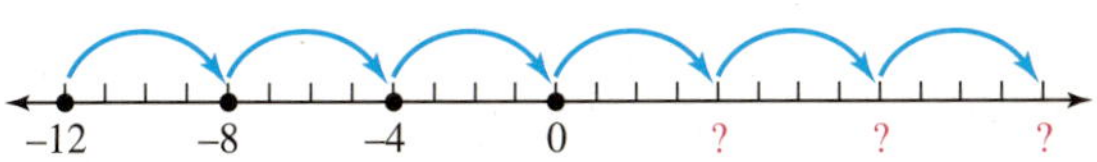

A graph of the products

From the pattern, we see that the product increases by 4 each time. Thus,

$$-4(-1) = 4, \qquad -4(-2) = 8, \qquad \text{and} \qquad -4(-3) = 12$$

These results illustrate that *the product of two negative integers is positive.* As a check, think of it as losing four debts of \$3. This is equivalent to gaining \$12. Therefore, $-4(-\$3) = \$12$.

We have seen that the product of two positive integers is positive, and the product of two negative integers is also positive. Those results illustrate the following rule.

Multiplying Two Integers That Have the Same (Like) Signs

To multiply two integers that have the same sign, multiply their absolute values. The final answer is positive.

Self Check 2

Multiply:

a. $-9(-7)$
b. $-12(-2)$
c. $-34(-15)$
d. $-4{,}100(-20{,}000)$

Now Try **Problems 33, 37, 41, and 43**

EXAMPLE 2 Multiply:

a. $-5(-9)$ **b.** $-8(-10)$ **c.** $-23(-42)$ **d.** $-2{,}500(-30{,}000)$

Strategy We will use the rule for multiplying two integers that have the same (like) signs.

WHY In each case, we are asked to multiply two negative integers.

Solution

a. Find the absolute values: $|-5| = 5$ and $|-9| = 9$.

$-5(-9) = 45$ — Multiply the absolute values, 5 and 9, to get 45. The final answer is positive.

b. Find the absolute values: $|-8| = 8$ and $|-10| = 10$.

$-8(-10) = 80$ — Multiply the absolute values, 8 and 10, to get 80. The final answer is positive.

c. Find the absolute values: $|-23| = 23$ and $|-42| = 42$.

$-23(-42) = 966$ — Multiply the absolute values, 23 and 42, to get 966. The final answer is positive.

$$\begin{array}{r} 42 \\ \times\ 23 \\ \hline 126 \\ 840 \\ \hline 966 \end{array}$$

d. We can extend the method for multiplying whole-number factors with trailing zeros to products of integers with trailing zeros.

$-2{,}500(-30{,}000) = 75{,}000{,}000$ — Attach six 0's after 75.

Multiply -25 and -3 to get 75.

We now summarize the multiplication rules for two integers.

Multiplying Two Integers

To multiply two nonzero integers, multiply their absolute values.

1. The product of two integers that have the same (*like*) signs is positive.
2. The product of two integers that have different (*unlike*) signs is negative.

Using Your CALCULATOR **Multiplication with Negative Numbers**

At Thanksgiving time, a large supermarket chain offered customers a free turkey with every grocery purchase of \$200 or more. Each turkey cost the store \$8, and 10,976 people took advantage of the offer. Since each of the 10,976 turkeys given away represented a loss of \$8 (which can be expressed as −\$8), the company lost a total of 10,976(−\$8). To perform this multiplication using a calculator, we enter the following:

Reverse entry: 10976 [×] 8 [+/−] [=] → −87808

Direct entry: 10976 [×] [(−)] 8 [ENTER] → −87808

The negative result indicates that with the turkey giveaway promotion, the supermarket chain lost \$87,808.

3 Perform several multiplications to evaluate expressions.

To evaluate expressions that contain several multiplications, we make repeated use of the rules for multiplying two integers.

EXAMPLE 3 Evaluate each expression:

a. $6(-2)(-7)$ **b.** $-9(8)(-1)$ **c.** $-3(-5)(2)(-4)$

Strategy Since there are no calculations within parentheses and no exponential expressions, we will perform the multiplications, working from the left to the right.

WHY This is the next step in the order of operations.

Solution

a. $6(-2)(-7) = -12(-7)$ Use the rule for multiplying two integers that have different signs: $6(-2) = -12$.

$= 84$ Use the rule for multiplying two integers that have the same sign.

$$\begin{array}{r} \overset{1}{1}2 \\ \times 7 \\ \hline 84 \end{array}$$

b. $-9(8)(-1) = -72(-1)$ Use the rule for multiplying two integers that have different signs: $-9(8) = -72$.

$= 72$ Use the rule for multiplying two integers that have the same sign.

c. $-3(-5)(2)(-4) = 15(2)(-4)$ Use the rule for multiplying two integers that have the same sign: $-3(-5) = 15$.

$= 30(-4)$ Use the rule for multiplying two integers that have the same sign: $15(2) = 30$.

$= -120$ Use the rule for multiplying two integers that have different signs.

Self Check 3

Evaluate each expression:

a. $3(-12)(-2)$

b. $-1(9)(-6)$

c. $-4(-5)(8)(-3)$

Now Try **Problems 45, 47, and 49**

In Section 1.2, we studied several properties of addition. A similar set of properties are true for multiplication.

Properties of Multiplication

Commutative property of multiplication: The order in which integers are multiplied does not change their product.

Associative property of multiplication: The way in which integers are grouped does not change their product.

Multiplication property of 0: The product of any integer and 0 is 0.

Multiplication property of 1: The product of any integer and 1 is that integer.

Another approach to evaluate expressions like those in Example 3 is to use the properties of multiplication to reorder and regroup the factors in a helpful way.

Self Check 4

Use the commutative and/or associative properties of multiplication to evaluate each expression from Self Check 3 in a different way:

a. $3(-12)(-2)$

b. $-1(9)(-6)$

c. $-4(-5)(8)(-3)$

Now Try **Problems 45, 47, and 49**

EXAMPLE 4 Use the commutative and/or associative properties of multiplication to evaluate each expression from Example 3 in a different way:

a. $6(-2)(-7)$ **b.** $-9(8)(-1)$ **c.** $-3(-5)(2)(-4)$

Strategy When possible, we will use the commutative and/or associative properties of multiplication to multiply pairs of negative factors.

WHY The product of two negative factors is positive. With this approach, we work with fewer negative numbers, and that lessens the possibility of an error.

Solution

a. $6(-2)(-7) = 6(14)$ Multiply the last two negative factors to produce a positive product: $-7(-2) = 14$.

$= 84$

$$\begin{array}{r} \overset{2}{1}4 \\ \times 6 \\ \hline 84 \end{array}$$

b. $-9(8)(-1) = 9(8)$ Multiply the negative factors to produce a positive product: $-9(-1) = 9$.

$= 72$

c. $-3(-5)(2)(-4) = 15(-8)$ Multiply the first two negative factors to produce a positive product. Multiply the last two factors.

$= -120$ Use the rule for multiplying two integers that have different signs.

$$\begin{array}{r} \overset{4}{1}5 \\ \times 8 \\ \hline 120 \end{array}$$

Self Check 5

Evaluate each expression:

a. $-1(-2)(-5)$

b. $-2(-7)(-1)(-2)$

Now Try **Problems 53 and 57**

EXAMPLE 5 Evaluate: **a.** $-2(-4)(-5)$ **b.** $-3(-2)(-6)(-5)$

Strategy When possible, we will use the commutative and/or associative properties of multiplication to multiply pairs of negative factors.

WHY The product of two negative factors is positive. With this approach, we work with fewer negative numbers, and that lessens the possibility of an error.

Solution

a. Note that this expression is the product of three (an odd number) negative integers.

$-2(-4)(-5) = 8(-5)$ Multiply the first two negative factors to produce a positive product.

$= -40$ The product is negative.

b. Note that this expression is the product of four (an even number) negative integers.

$-3(-2)(-6)(-5) = 6(30)$ Multiply the first two negative factors and the last two negative factors to produce positive products.

$= 180$ The product is positive.

Example 5, part a, illustrates that a product is negative when there is an odd number of negative factors. Example 5, part b, illustrates that a product is positive when there is an even number of negative factors.

Multiplying an Even and an Odd Number of Negative Integers

The product of an even number of negative integers is positive.
The product of an odd number of negative integers is negative.

4 Solve application problems by multiplying integers.

Problems that involve repeated addition are often more easily solved using multiplication.

EXAMPLE 6 *Oceanography*

Scientists lowered an underwater vessel called a *submersible* into the Pacific Ocean to record the water temperature. The first measurement was made 75 feet below sea level, and more were made every 75 feet until it reached the ocean floor. Find the depth of the submersible when the 25th measurement was made.

Emory Kristof/National Geographic/Getty Images

Solution It is helpful to list what we know and what we are to find.

- The first measurement was made 75 feet below sea level. *Given*
- More measurements were made every 75 feet. *Given*
- Find the depth of the submersible when it made the 25th measurement. *Find*

If we use negative numbers to represent the depths at which the measurements were made, then the first was at −75 feet. The depth (in feet) of the submersible when the 25th measurement was made can be found by adding −75 twenty-five times. This repeated addition can be calculated more simply by multiplication.

We translate the words of the problem to numbers and symbols.

The depth of the submersible when it made the 25th measurement	is equal to	the number of measurements made	times	the amount it was lowered each time.
The depth of the submersible when it made the 25th measurement	=	25	·	(−75)

To find the product, we use the rule for multiplying two integers that have different signs. First, we find the absolute values: $|25| = 25$ and $|-75| = 75$.

$25(-75) = -1{,}875$ Multiply the absolute values, 25 and 75, to get 1,875. Since the integers have different signs, make the final answer negative.

$$\begin{array}{r} 75 \\ \times\ 25 \\ \hline 375 \\ 1\ 500 \\ \hline 1{,}875 \end{array}$$

The depth of the submersible was 1,875 feet below sea level (−1,875 feet) when the 25th temperature measurement was taken.

We can use estimation or simply perform the actual multiplication again to see if the result seems reasonable.

Self Check 6

GASOLINE LEAKS To determine how badly a gasoline tank was leaking, inspectors used a drilling process to take soil samples nearby. The first sample was taken 6 feet below ground level, and more were taken every 6 feet after that. The 14th sample was the first one that did not show signs of gasoline. How far below ground level was that?

Now Try **Problem 95**

ANSWERS TO SELF CHECKS

1. a. −12 **b.** −120 **c.** −1,275 **d.** −98,000 **2. a.** 63 **b.** 24 **c.** 510 **d.** 82,000,000 **3. a.** 72 **b.** 54 **c.** −480 **4. a.** 72 **b.** 54 **c.** −480 **5. a.** −10 **b.** 28 **6.** 84 ft below ground level (−84 ft)

SECTION 1.4 STUDY SET

VOCABULARY

Fill in the blanks.

1. In the multiplication problem shown below, label each *factor* and the *product.*

$$-5 \cdot 10 = -50$$

(blank) (blank) (blank)

2. Two negative integers, as well as two positive integers, are said to have the same signs or _____ signs.

3. A positive integer and a negative integer are said to have different signs or _______ signs.

4. ____________ property of multiplication: The order in which integers are multiplied does not change their product.

5. __________ property of multiplication: The way in which integers are grouped does not change their product.

6. Multiplication is repeated ________.

CONCEPTS

Fill in the blanks.

7. Multiplication of integers is very much like multiplication of whole numbers. The only difference is that we must determine whether the answer is positive or _________.

8. When we multiply two nonzero integers, they either have different signs or ______ sign.

9. To multiply a positive integer and a negative integer, multiply their absolute values. Then make the final answer _________.

10. To multiply two integers that have the same sign, multiply their absolute values. The final answer is ________.

11. The product of two integers with _____________ signs is negative.

12. The product of two integers with ___________ signs is positive.

13. The product of any integer and 0 is ___ .

14. The product of an even number of negative integers is ________ and the product of an odd number of negative integers is _________.

15. Find each absolute value.

a. $|-3|$ **b.** $|12|$

16. Which property justifies each statement?

a. $-5(2 \cdot 17) = (-5 \cdot 2)17$

b. $-5\left(-\frac{1}{5}\right) = 1$

c. $-5 \cdot 2 = 2(-5)$

d. $-5(1) = -5$

e. $-5 \cdot 0 = 0$

NOTATION

17. Translate to mathematical symbols.

a. negative three times negative two

b. five times negative five

c. the opposite of five times five

18. The product of negative four and negative five is twenty.

19. Complete each statement using the given property.

a. $5 \cdot 8 =$ ___ Commutative property of multiplication

b. $-2(6 \cdot 9) =$ ___ Associative property of multiplication

c. $5 \times$ ___ $= 0$ Multiplication property of zero.

d. ___ $(-20) = -20$ Multiplication property of 1

Complete the steps to evaluate the expression.

20. $-3(-2)(-4) =$ ___ (-4)

$=$ ___

GUIDED PRACTICE

Multiply. See Example 1.

21. $5(-3)$ **22.** $4(-6)$

23. $9(-2)$ **24.** $5(-7)$

25. $18(-4)$ **26.** $17(-8)$

27. $21(-6)$ **28.** $39(-3)$

29. $-45 \cdot 37$ **30.** $-42 \cdot 24$

31. $-94 \cdot 1{,}000$ **32.** $-76 \cdot 1{,}000$

Multiply. See Example 2.

33. $(-8)(-7)$ **34.** $(-9)(-3)$

35. $-7(-1)$ **36.** $-5(-1)$

37. $-3(-52)$ **38.** $-4(-73)$

39. $-6(-46)$ **40.** $-8(-48)$

41. $-59(-33)$ **42.** $-61(-29)$

43. $-60{,}000(-1{,}200)$ **44.** $-20{,}000(-3{,}200)$

Evaluate each expression. See Examples 3 and 4.

45. $6(-3)(-5)$ **46.** $9(-3)(-4)$

47. $-5(10)(-3)$ **48.** $-8(7)(-2)$

49. $-2(-4)(6)(-8)$ **50.** $-3(-5)(2)(-9)$

51. $-8(-3)(7)(-2)$ **52.** $-9(-3)(4)(-2)$

Evaluate each expression. See Example 5.

53. $-4(-2)(-6)$

54. $-4(-6)(-3)$

55. $-3(-9)(-3)$

56. $-5(-2)(-5)$

57. $-1(-3)(-2)(-6)$

58. $-1(-4)(-2)(-4)$

59. $-9(-4)(-1)(-4)$

60. $-6(-3)(-6)(-1)$

TRY IT YOURSELF

Evaluate each expression.

61. $6(-5)(2)$

62. $4(-2)(2)$

63. $-8(0)$

64. $0(-27)$

65. $63(-7)$

66. $43(-6)$

67. $(-2)10$

68. $(-3)8$

69. $-2(-3)(3)(-1)$

70. $5(-2)(3)(-1)$

71. Find the product of -6 and the opposite of 10.

72. Find the product of the opposite of 9 and the opposite of 8.

73. $-6(-4)(-2)$

74. $-3(-2)(-3)$

75. $-42 \cdot 200{,}000$

76. $-56 \cdot 10{,}000$

77. $-3 \cdot 4$

78. $(-9)(11)$

79. $-12(-12)$

80. $-5(-5)$

81. $(-6)(-9)$

82. $(-8)(-7)$

83. $(-1)(-2)(-3)(-4)(-5)$

84. $(-10)(-8)(-6)(-4)(-2)$

85. $(-6)(-6)(-6)$

86. $(-5)(-5)(-5)$

87. $-3(-4)(0)$

88. $15(0)(-22)$

CONCEPT EXTENSIONS

In problems 89 and 90, write a number sentence of the form ☐(☐) = ☐ for each situation.

89. **a.** The product of two integers is -54.

b. The product of two negative integers is 54.

90. **a.** The product of two integers is -115.

b. The product of two negative integers is 115.

91. Find three integers whose product is negative.

92. Find two integers whose sum is -10 and whose product is -24.

93. Find two integers whose sum is -9 and whose product is 20.

94. Refer to the number line below, where a, b, c, and d represent integers. Determine whether each product is positive or negative.

a. $a \cdot b$

b. $a \cdot c$

c. $a \cdot d \cdot b$

d. $a \cdot b \cdot c \cdot d$

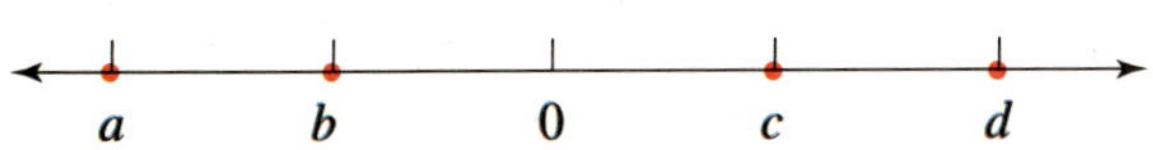

APPLICATIONS

Use signed numbers to solve each problem.

95. SUBMARINES As part of a training exercise, the captain of a submarine ordered it to descend 250 feet, level off for 5 minutes, and then repeat the process several times. If the sub was on the ocean's surface at the beginning of the exercise, find its depth after the 8th dive.

96. BUILDING A PIER A *pile driver* uses a heavy weight to pound tall poles into the ocean floor. If each strike of a pile driver on the top of a pole sends it 6 inches deeper, find the depth of the pole after 20 strikes.

Image Source/Getty Images

97. MAGNIFICATION A mechanic used an electronic testing device to check the smog emissions of a car. The results of the test are displayed on a screen.

a. Find the high and low values for this test as shown on the screen.

b. By switching a setting, the picture on the screen can be magnified. What would be the new high and new low if every value were doubled?

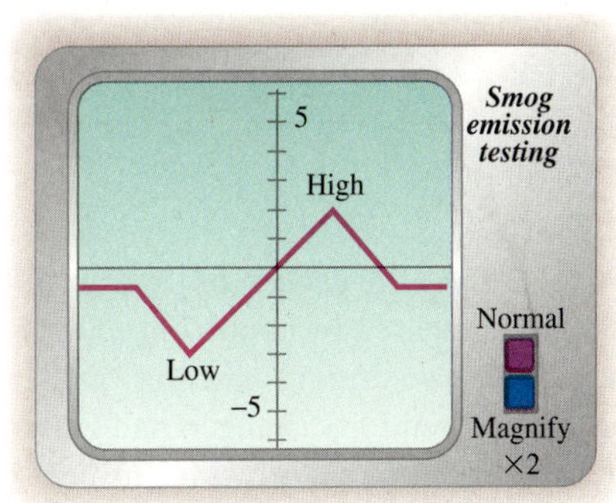

98. LIGHT Water acts as a selective filter of light. In the illustration, we see that red light waves penetrate water only to a depth of about 5 meters. See diagram on the next page. How many times deeper does

a. yellow light penetrate than red light?

b. green light penetrate than orange light?

c. blue light penetrate than yellow light?

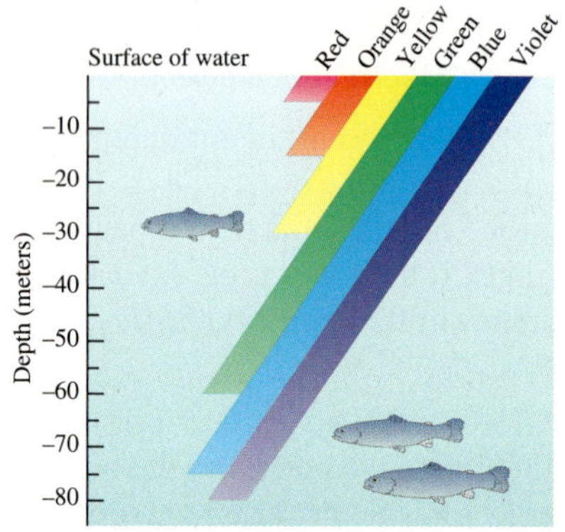

99. RECESSIONS Refer to the bar graph. The year 2008 was one of the worst in U.S. history for job losses. Find the number of jobs lost in . . .

a. September 2008 if it was about 6 times the number lost in April.

b. October 2008 if it was about 9 times the number lost in May.

c. November 2008 if it was about 7 times the number lost in February.

d. December if it was about 6 times the number lost in March.

Source: Bureau of Labor Statistics

100. RUSSIA During the decade of the 1990's, Russia population decreased by about 700,000 per year because of high death rates and low birth rates. Recently, that trend has changed and Russia is now experiencing population growth. Estimate the total decline in Russia's population for the decade of the 1990's? (Source: Russian FederalService of State Statistics)

101. PLANETS The average surface temperature of Mars is $-81°F$. Find the average surface temperature of Uranus if it is four times colder than Mars. (Source: *The World Almanac and Book of Facts*, 2009)

102. CROP LOSS A farmer, worried about his fruit trees suffering frost damage, calls the weather service for temperature information. He is told that temperatures will be decreasing approximately 5 degrees every hour for the next five hours. What signed number represents the total change in temperature expected over the next five hours?

103. TAX WRITE-OFF For each of the last six years, a businesswoman has filed a $200 depreciation allowance on her income tax return for an office computer system. What signed number represents the total amount of depreciation written off over the six-year period?

104. EROSION A levee protects a town in a low-lying area from flooding. According to geologists, the banks of the levee are eroding at a rate of 2 feet per year. If something isn't done to correct the problem, what signed number indicates how much of the levee will erode during the next decade?

105. DECK SUPPORTS After a winter storm, a homeowner has an engineering firm inspect his damaged deck. Their report concludes that the original foundation poles were not sunk deep enough, by a factor of 3. What signed number represents the depth to which the poles should have been sunk?

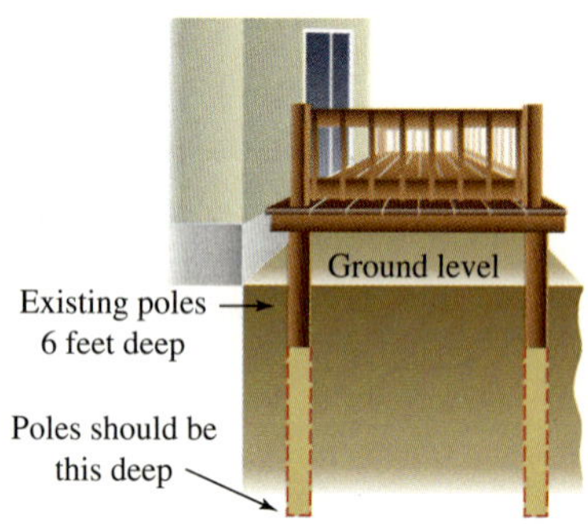

106. DIETING After giving a patient a physical exam, a physician felt that the patient should begin a diet. The two options that were discussed are shown in the following table.

	Plan #1	Plan #2
Length	10 weeks	14 weeks
Daily exercise	1 hr	30 min
Weight loss per week	3 lb	2 lb

a. Find the expected weight loss from Plan 1. Express the answer as a signed number.

b. Find the expected weight loss from Plan 2. Express the answer as a signed number.

c. With which plan should the patient expect to lose more weight? Explain why the patient might not choose it.

107. ADVERTISING The paid attendance for the last night of the 2008 Rodeo Houston was 71,906. Suppose a local country music radio station gave a sports bag, worth \$3, to everyone that attended. Find the signed number that expresses the radio station's financial loss from this giveaway.

108. HEALTH CARE A health care provider for a company estimates that 75 hours per week are lost by employees suffering from stress-related or preventable illness. In a 52-week year, how many hours are lost? Use a signed number to answer.

109. FLUID FLOW In a lab, the temperature of a fluid was decreased 6° per hour for 12 hours. What signed number indicates the change in temperature?

110. ASTRONOMY The temperature on Pluto gets as low as −386°F. This is twice as low as the lowest temperature reached on Jupiter. What is the lowest temperature on Jupiter?

111. COMPUTERS The formula = A1*B1*C1 in cell D1 of the spreadsheet instructs the computer to multiply the values in cells A1, B1, and C1 and to print the result *in place of the formula* in cell D1. (The symbol * represents multiplication.) What value will be printed in the cell D1? What values will be printed in cells D2 and D3?

File Edit View Insert Format Tools Data Window

	A	B	C	D
1	4	-5	-17	= A1•B1•C1
2	22	-30	14	= A2•B2•C2
3	-60	-20	-34	= A3•B3•C3
4				
5				

112. WEIGHT LOSS. As a result of a diet, Tom has been steadily losing 4 pounds per month.

a. Which expression below can be used to determine how much heavier Tom was 8 months ago?

i. $-4 \cdot 8$ **ii.** $-4(-8)$

iii. $4(-8)$ **iv.** $-4 - 8$

b. How much heavier was Tom 8 months ago?

WRITING

113. Explain why the product of a positive number and a negative number is negative, using $5(-3)$ as an example.

114. Explain the multiplication rule for integers that is shown in the pattern of signs below.

$$(-)(-) = +$$
$$(-)(-)(-) = -$$
$$(-)(-)(-)(-) = +$$
$$(-)(-)(-)(-)(-) = -$$
$$\vdots$$

115. When a number is multiplied by −1, the result is the opposite of the original number. Explain why.

116. A student claimed, "A positive and a negative is negative." What is wrong with this statement?

117. The commutative property states that changing the order when multiplying does not change the answer. Are the following activities commutative? Explain.

a. Washing a load of clothes; drying a load of clothes

b. Putting on your left sock; putting on your right sock

118. If we multiply two different numbers and the answer is 0, what must be true about one of the numbers? Explain your answer.

SECTION 1.5 Dividing Integers

Objectives

1. Divide two integers.
2. Identify *division of 0* and *division by 0.*
3. Solve application problems by dividing integers.

ARE YOU READY?

The following problems review some basic concepts that are important when dividing integers.

1. Find $|-45|$ and $|225|$.
2. Do the integers 27 and −9 have the same sign or different signs?
3. Divide: $72 \div 8$
4. Divide: $\frac{900}{10}$
5. Divide: $9\overline{)1{,}962}$
6. Multiply: $(-8)(-7)$

In this section, we will develop rules for division of integers, just as we did earlier for multiplication of integers.

Division problems can be written using a **division symbol** ÷, a **long division symbol** $\overline{)\quad}$, or a **fraction bar** —. We call the number being divided the **dividend** and the number that we are dividing by is called the **divisor.** The answer is called the **quotient.**

Division symbol	Long division symbol	Fraction bar
$6 \div 3 = 2$ (6: Dividend; 3: Divisor; 2: Quotient)	$3\overline{)6}$ with 2 on top (2: Quotient; 3: Divisor; 6: Dividend)	$\frac{6}{3} = 2$ (6: Dividend; 3: Divisor; 2: Quotient)

1 Divide two integers.

Every division has a related multiplication statement. For example,

$$\frac{6}{3} = 2 \qquad \text{because} \qquad 2(3) = 6$$

and

$$\frac{20}{5} = 4 \qquad \text{because} \qquad 4(5) = 20$$

We can use the relationship between multiplication and division to help develop rules for dividing integers. There are four cases to consider.

Case 1: A positive integer divided by a positive integer
From years of experience, we already know that the result is positive. Therefore, *the quotient of two positive integers is positive.*

Case 2: A negative integer divided by a negative integer
As an example, consider the division $\frac{-12}{-2} = ?$. We can find ? by examining the related multiplication statement.

Related multiplication statement	Division statement
$?(-2) = -12$	$\frac{-12}{-2} = ?$
This must be positive 6 if the product is to be negative 12.	So the quotient is positive 6.

Therefore, $\frac{-12}{-2} = 6$. This example illustrates that *the quotient of two negative integers is positive.*

Case 3: A positive integer divided by a negative integer
Let's consider $\frac{12}{-2} = ?$. We can find ? by examining the related multiplication statement.

Related multiplication statement	Division statement
$?(-2) = 12$	$\frac{12}{-2} = ?$
This must be -6 if the product is to be positive 12.	So the quotient is -6.

Therefore, $\frac{12}{-2} = -6$. This example illustrates that *the quotient of a positive integer and a negative integer is negative.*

Case 4: A negative integer divided by a positive integer
Let's consider $\frac{-12}{2} = ?$. We can find ? by examining the related multiplication statement.

Related multiplication statement

$?(2) = -12$

This must be -6 if the product is to be -12.

Division statement

$\frac{-12}{2} = ?$

So the quotient is -6.

Therefore, $\frac{-12}{2} = -6$. This example illustrates that *the quotient of a negative integer and a positive integer is negative.*

We now summarize the results from the previous examples and note that they are similar to the rules for multiplication.

Dividing Two Integers

To divide two integers, divide their absolute values.

1. The quotient of two integers that have the same (*like*) signs is positive.
2. The quotient of two integers that have different (*unlike*) signs is negative.

EXAMPLE 1

Divide and check the result:

a. $\frac{-14}{7}$ **b.** $30 \div (-5)$ **c.** $\frac{176}{-11}$ **d.** $-24{,}000 \div 600$

Strategy We will use the rule for dividing two integers that have different (unlike) signs.

WHY Each division involves a positive and a negative integer.

Solution

a. Find the absolute values: $|-14| = 14$ and $|7| = 7$.

$\frac{-14}{7} = -2$ Divide the absolute values, 14 by 7, to get 2. Then make the final answer negative.

To check, we multiply the *quotient*, -2, and the *divisor*, 7. We should get the *dividend*, -14.

Check: $-2(7) = -14$ The result checks.

b. Find the absolute values: $|30| = 30$ and $|-5| = 5$.

$30 \div (-5) = -6$ Divide the absolute values, 30 by 5, to get 6. Then make the final answer negative.

Check: $-6(-5) = 30$ The result checks.

c. Find the absolute values: $|176| = 176$ and $|-11| = 11$.

$\frac{176}{-11} = -16$ Divide the absolute values, 176 by 11, to get 16. Then make the final answer negative.

$$\begin{array}{r} 16 \\ 11\overline{)176} \\ -11 \\ \hline 66 \\ -66 \\ \hline 0 \end{array}$$

Check: $-16(-11) = 176$ The result checks.

d. If a divisor has ending zeros, we can simplify the division by removing the same number of ending zeros in the divisor and dividend.

There are two zeros in the divisor.

$-24{,}000 \div 600 = -240 \div 6 = -40$

Remove two zeros from the dividend and the divisor, and divide.

Divide the absolute values, 240 by 6, to get 40. Then make the final answer negative.

Check: $-40(600) = -24{,}000$ Use the original divisor and dividend in the check.

Self Check 1

Divide and check the result:

a. $\frac{-45}{5}$

b. $28 \div (-4)$

c. $\frac{336}{-14}$

d. $-18{,}000 \div 300$

Now Try **Problems 13, 15, 21, and 27**

Caution! When dividing two numbers it is important that we write them in the correct order, because ***division is not commutative.*** For instance, in Example 1 part a, note that $\frac{-14}{7} \neq \frac{7}{-14}$. In fact, $\frac{7}{-14}$ is not even an integer.

Self Check 2

Divide and check the result:

a. $\frac{-27}{-3}$

b. $-24 \div (-4)$

c. $\frac{-301}{-7}$

d. $-400 \div (-20)$

Now Try **Problems 33, 37, 41, and 43**

EXAMPLE 2 Divide and check the result:

a. $\frac{-12}{-3}$ **b.** $-48 \div (-6)$ **c.** $\frac{-315}{-9}$ **d.** $-200 \div (-40)$

Strategy We will use the rule for dividing two integers that have the same (like) signs.

WHY In each case, we are asked to find the quotient of two negative integers.

Solution

a. Find the absolute values: $|-12| = 12$ and $|-3| = 3$.

$\frac{-12}{-3} = 4$ Divide the absolute values, 12 by 3, to get 4. The final answer is positive.

Check: $4(-3) = -12$ The result checks.

b. Find the absolute values: $|-48| = 48$ and $|-6| = 6$.

$-48 \div (-6) = 8$ Divide the absolute values, 48 by 6, to get 8. The final answer is positive.

Check: $8(-6) = -48$ The result checks.

c. Find the absolute values: $|-315| = 315$ and $|-9| = 9$.

$\frac{-315}{-9} = 35$ Divide the absolute values, 315 by 9, to get 35. The final answer is positive.

$$\begin{array}{r} 35 \\ 9\overline{)315} \\ -27 \\ \hline 45 \\ -45 \\ \hline 0 \end{array}$$

Check: $35(-9) = -315$ The result checks.

d. We can simplify the division by removing the same number of ending zeros in the divisor and dividend.

There is one zero in the divisor.

$-200 \div (-40) = -20 \div (-4) = 5$ Divide the absolute values, 20 by 4, to get 5. The final answer is positive.

Remove one zero from the dividend and the divisor, and divide.

Check: $5(-40) = -200$ The result checks.

2 Identify *division of 0* and *division by 0.*

To review the concept of division of 0, we consider $\frac{0}{-2} = ?$. We can attempt to find ? by examining the related multiplication statement.

Related multiplication statement	**Division statement**
$(?)(-2) = 0$ — This must be 0 if the product is to be 0.	$\frac{0}{-2} = ?$ — So the quotient is 0.

Therefore, $\frac{0}{-2} = 0$. This example illustrates that *the quotient of 0 divided by any nonzero integer is 0.*

To review division by 0, let's consider $\frac{-2}{0} = ?$. We can attempt to find ? by examining the related multiplication statement.

Related multiplication statement	**Division statement**
$(?)0 = -2$ — There is no number that gives -2 when multiplied by 0.	$\frac{-2}{0} = ?$ — There is no quotient.

Therefore, $\frac{-2}{0}$ does not have an answer and we say that $\frac{-2}{0}$ is undefined. This example illustrates that *the quotient of any nonzero integer divided by 0 is undefined.*

Division with 0

1. If 0 is divided by any nonzero integer, the quotient is 0.
2. Division of any nonzero integer by 0 is undefined.

EXAMPLE 3 Divide, if possible: **a.** $\frac{-4}{0}$ **b.** $0 \div (-8)$

Strategy In each case, we need to determine if we have division *of* 0 or division *by* 0.

WHY *Division of 0* by a nonzero integer is defined, and the answer is 0. However, *division of a nonzero integer by 0* is undefined; there is no answer.

Solution

a. $\frac{-4}{0}$ is undefined. This is division by 0.

b. $0 \div (-8) = 0$ because $0(-8) = 0$. This is division of 0.

Self Check 3

Divide, if possible:

a. $\frac{-12}{0}$ **b.** $0 \div (-6)$

Now Try Problems 45 and 47

3 Solve application problems by dividing integers.

Problems that involve forming equal-sized groups can be solved by division.

EXAMPLE 4 ***Real Estate*** Over the course of a year, a homeowner reduced the price of his house by an equal amount each month, because it was not selling. By the end of the year, the price was \$11,400 less than at the beginning of the year. By how much was the price of the house reduced each month?

David McNew/Getty Images

Analyze

- The homeowner dropped the price \$11,400 in 1 year. *Given*
- The price was reduced by an equal amount each month. *Given*
- By how much was the price of the house reduced each month? *Find*

We can express the drop in the price of the house for the year as $-\$11,400$. The phrase *reduced by an equal amount each month* indicates division.

We translate the words of the problem to numbers and symbols.

Self Check 4

SELLING BOATS The owner of a sail boat reduced the price of the boat by an equal amount each month, because there were no interested buyers. After 8 months, and a \$960 reduction in price, the boat sold. By how much was the price of the boat reduced each month?

Now Try Problem 87

The amount the price was reduced each month	is equal to	the drop in the price of the house for the year	divided by	the number of months in 1 year.
The amount the price was reduced each month	=	−11,400	÷	12

To find the quotient, we use the rule for dividing two integers that have different signs. First, we find the absolute values: $|-11{,}400| = 11{,}400$ and $|12| = 12$.

$-11{,}400 \div 12 = -950$ Divide the absolute values, 11,400 and 12, to get 950. Then make the final answer negative.

```
    950
12)11,400
  - 10 8
      60
    - 60
      00
    - 00
       0
```

The negative result indicates that the price of the house was *reduced* by $950 each month.

We can use estimation to check the result. A reduction of $1,000 each month would cause the price to drop $12,000 in 1 year. It seems reasonable that a reduction of $950 each month would cause the price to drop $11,400 in a year.

The Language of Algebra Depreciation is a form of the word *depreciate,* meaning to lose value. You've probably heard that the minute you drive a new car off the lot, it has *depreciated.*

Self Check 5

DEPRECIATION Over a 6-year period, the value of a $300,000 house fell at a uniform rate to $286,500. Find the amount of depreciation each year.

Now Try **Problem 99**

EXAMPLE 5 ***Depreciation.*** Over an 8-year period, the value of a $150,000 house fell at a uniform rate to $110,000. Find the amount of depreciation per year.

©iStockPhoto.com/EyeMark

Strategy The phrase *uniform rate* means that the value of the house fell the same amount each year, for 8 straight years. We can determine the amount it depreciated in one year (per year) by dividing the total change in value of the house by 8.

WHY The process of separating a quantity into equal parts (in this case, the change in the value of the house) indicates division.

Solution

First, we find the change in the value of the house.

$110{,}000 - 150{,}000 = -40{,}000$ Subtract the previous value from the current value.

The negative result represents a drop in value of $40,000. Since the depreciation occurred over 8 years, we divide −40,000 by 8.

$\frac{-40{,}000}{8} = -5{,}000$ Divide the absolute values, 40,000 by 8, to get 5,000, and make the quotient negative.

The house depreciated $5,000 per year.

Using Your CALCULATOR Division with Negative Numbers

The Bureau of Labor Statistics estimated that the United States lost 162,000 auto manufacturing jobs (motor vehicles and parts) in 2008. Because the jobs were lost, we write this as $-162{,}000$. To find the average number of manufacturing jobs lost each month, we divide: $\frac{-162{,}000}{12}$. We can use a calculator to perform the division.

Reverse entry: 162000 [+/−] [÷] 12 [=]

Direct entry: 162000 [÷] [(−)] 12 [ENTER] [−13500]

The average number of auto manufacturing jobs lost each month in 2008 was 13,500.

ANSWERS TO SELF CHECKS

1. a. -9 **b.** -7 **c.** -24 **d.** -60 **2. a.** 9 **b.** 6 **c.** 43 **d.** 20 **3. a.** undefined **b.** 0 **4.** The price was reduced by \$120 each month. **5.** The house depreciated \$2,250 per year.

SECTION 1.5 STUDY SET

VOCABULARY

Fill in the blanks.

1. In the division problems shown below, label the *dividend, divisor,* and *quotient.*

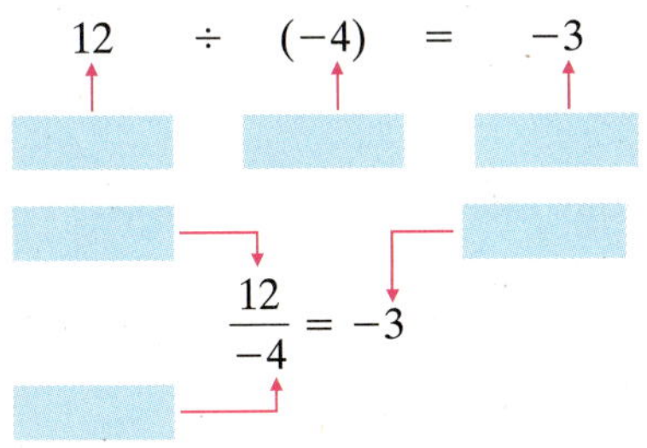

2. The related ___________ statement for $\frac{-6}{3} = -2$ is $-2(3) = -6$.

3. $\frac{-3}{0}$ is division ____ 0 and $\frac{0}{-3} = 0$ is division ____ 0.

4. Division of a nonzero integer by 0, such as $\frac{-3}{0}$, is ________.

CONCEPTS

5. Write the related multiplication statement for each division.

a. $\frac{-25}{5} = -5$ **b.** $-36 \div (-6) = 6$ **c.** $\frac{0}{-15} = 0$

6. Using multiplication, check to determine whether $-720 \div 45 = -12$.

7. Fill in the blanks.
To divide two integers, divide their absolute values.

a. The quotient of two integers that have the same (*like*) signs is _______.

b. The quotient of two integers that have different (*unlike*) signs is _______.

8. If a divisor has ending zeros, we can simplify the division by removing the same number of ending zeros in the divisor and dividend. Fill in the blank: $-2{,}400 \div 60 = -240 \div$ ☐

9. Fill in the blanks.

a. If 0 is divided by any nonzero integer, the quotient is ☐.

b. Division of any nonzero integer by 0 is _________.

10. What operation can be used to solve problems that involve forming equal-sized groups?

11. Determine whether each statement is always true, sometimes true, or never true.

a. The product of a positive integer and a negative integer is negative.

b. The sum of a positive integer and a negative integer is negative.

c. The quotient of a positive integer and a negative integer is negative.

12. Determine whether each statement is always true, sometimes true, or never true.

a. The product of two negative integers is positive.

b. The sum of two negative integers is negative.

c. The quotient of two negative integers is negative.

GUIDED PRACTICE

Divide and check the result. See Example 1.

13. $\frac{-14}{2}$

14. $\frac{-10}{5}$

15. $\frac{-20}{5}$

16. $\frac{-24}{3}$

17. $36 \div (-6)$

18. $36 \div (-9)$

19. $24 \div (-3)$

20. $42 \div (-6)$

21. $\frac{264}{-12}$

22. $\frac{364}{-14}$

23. $\frac{702}{-18}$

24. $\frac{396}{-12}$

25. $-9{,}000 \div 300$

26. $-12{,}000 \div 600$

27. $-250{,}000 \div 5{,}000$

28. $-420{,}000 \div 7{,}000$

Divide and check the result. See Example 2.

29. $\frac{-8}{-4}$

30. $\frac{-12}{-4}$

31. $\frac{-45}{-9}$

32. $\frac{-81}{-9}$

33. $-63 \div (-7)$

34. $-21 \div (-3)$

35. $-32 \div (-8)$

36. $-56 \div (-7)$

37. $\frac{-400}{-25}$

38. $\frac{-490}{-35}$

39. $\frac{-651}{-31}$

40. $\frac{-736}{-32}$

41. $-800 \div (-20)$

42. $-800 \div (-40)$

43. $-15{,}000 \div (-30)$

44. $-36{,}000 \div (-60)$

Divide, if possible. See Example 3.

45. a. $\frac{-3}{0}$ **b.** $\frac{0}{-3}$

46. a. $\frac{-5}{0}$ **b.** $\frac{0}{-5}$

47. a. $\frac{0}{-24}$ **b.** $\frac{-24}{0}$

48. a. $\frac{0}{-32}$ **b.** $\frac{-32}{0}$

TRY IT YOURSELF

Divide, if possible.

49. $-36 \div (-12)$

50. $-45 \div (-15)$

51. $\frac{425}{-25}$

52. $\frac{462}{-42}$

53. $0 \div (-16)$

54. $0 \div (-6)$

55. Find the quotient of -45 and 9.

56. Find the quotient of -36 and -4.

57. $-2{,}500 \div 500$

58. $-52{,}000 \div 4{,}000$

59. $\frac{-6}{0}$

60. $\frac{-8}{0}$

61. $\frac{-19}{1}$

62. $\frac{-9}{1}$

63. $-23 \div (-23)$

64. $-11 \div (-11)$

65. $\frac{40}{-2}$

66. $\frac{35}{-7}$

67. $9 \div (-9)$

68. $15 \div (-15)$

69. $\frac{-10}{-1}$

70. $\frac{-12}{-1}$

71. $\frac{-888}{37}$

72. $\frac{-456}{24}$

73. $\frac{3{,}000}{-100}$

74. $\frac{-60{,}000}{-1{,}000}$

75. Divide 8 by -2.

76. Divide -16 by -8.

LOOK ALIKES . . .

77. a. $27 + (-9)$ **b.** $27 - (-9)$

c. $27(-9)$ **d.** $\frac{27}{-9}$

78. a. $-55 + (-11)$ **b.** $-55 - (-11)$

c. $-55(-11)$ **d.** $\frac{-55}{-11}$

CONCEPT EXTENSIONS

In problems 79-81, write a number sentence of the form $\frac{\square}{\square} = \square$ *for each situation.*

79. a. The quotient of two negative integers is 15.

b. The quotient of two integers is -15.

80. a. The quotient of two integers is 0.

b. The quotient of two integers is undefined.

81. a. The quotient of two negative integers is 1.

b. The quotient of two integers is -1.

82. *Let POS stand for a positive number and NEG stand for a negative number. Determine the sign of each result, if possible.*

a. POS · NEG

b. POS + NEG

c. POS − NEG **d.** $\frac{\text{POS}}{\text{NEG}}$

e. NEG · NEG **f.** NEG + NEG

g. NEG − NEG

h. $\frac{\text{NEG}}{\text{NEG}}$

Use a calculator to perform each division.

83. $\frac{-13{,}550}{25}$

84. $\frac{3{,}876}{-19}$

85. $\frac{27{,}778}{-17}$

86. $\frac{-168{,}476}{-77}$

APPLICATIONS

Use signed numbers to solve each problem.

87. LOWERING PRICES A furniture store owner reduced the price of an oak table an equal amount each week, because it was not selling. After six weeks, and a $210 reduction in price, the table was purchased. By how much was the price of the table reduced each week?

88. TEMPERATURE DROP During a five-hour period, the temperature steadily dropped 20°F. By how many degrees did the temperature change each hour?

89. SUBMARINES In a series of three equal dives, a submarine is programmed to reach a depth of 3,030 feet below the ocean surface. What signed number describes how deep each of the dives will be?

90. GRAND CANYON A mule train is to travel from a stable on the rim of the Grand Canyon to a camp on the canyon floor, approximately 5,500 feet below the rim. If the guide wants the mules to be rested after every 500 feet of descent, how many stops will be made on the trip?

91. CHEMISTRY During an experiment, a solution was steadily chilled and the times and temperatures were recorded, as shown in the illustration below. By how many degrees did the temperature of the solution change each minute?

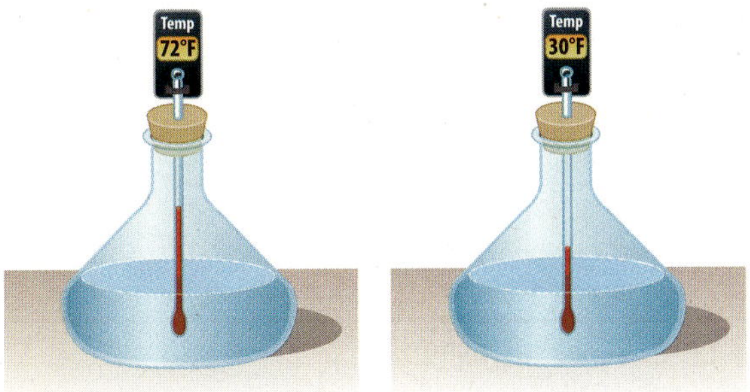

Beginning of experiment 8:00 A.M.   End of experiment 8:06 A.M.

92. OCEAN EXPLORATION The Mariana Trench is the deepest part of the world's oceans. It is located in the North Pacific Ocean near the Philippines and has a maximum depth of 36,201 feet. If a remote-controlled vessel is sent to the bottom of the trench in a series of 11 equal descents, how far will the vessel descend on each dive? (Source: marianatrench.com)

93. BASEBALL TRADES At the midway point of the season, a baseball team finds itself 12 games behind the league leader. Team management decides to trade for a talented hitter, in hopes of making up at least half of the deficit in the standings by the end of the year. Where in the league standings does management expect to finish at season's end?

94. BUDGET DEFICITS A politician proposed a two-year plan for cutting a county's $20-million budget deficit, as shown. If this plan is put into effect, how will the deficit change in two years?

	Plan	Prediction
1st year	Raise taxes, drop failing programs	Will cut deficit in half
2nd year	Search out waste and fraud	Will cut remaining deficit in half

95. MARKDOWNS The owner of a clothing store decides to reduce the price on a line of jeans that are not selling. She feels she can afford to lose $300 of projected income on these pants. By how much can she mark down each of the 20 pairs of jeans?

96. WATER STORAGE Over a week's time, engineers at a city water reservoir released enough water to lower the water level 105 feet. On average, how much did the water level change each day during this period?

97. THE STOCK MARKET On Monday, the value of Maria's 255 shares of stock was at an all-time high. By Friday, the value had fallen $4,335. What was her per-share loss that week?

98. CUTTING BUDGETS In a cost-cutting effort, a company decides to cut $5,840,000 from its annual budget. To do this, all of the company's 160 departments will have their budgets reduced by an equal amount. By how much will each department's budget be reduced?

99. REAL ESTATE Over a 5-year period, the value of a $200,000 lot fell at a uniform rate to $160,000. What signed number indicates the amount of depreciation per year?

100. TOURISM The ocean liner Queen Mary cost $22,500,000 to build in 1936. The ship was purchased by the city of Long Beach, California, in 1967 for $3,435,000. It now serves as a convention center.

a. What signed number indicates the amount of depreciation of the Queen Mary from 1936 to 1967?

b. What signed number indicates the annual average depreciation of the ship over the 31-year period from 1936 to 1967?

WRITING

101. Explain why the quotient of two negative integers is positive.

102. How do the rules for multiplying integers compare with the rules for dividing integers?

103. Use a specific example to explain how multiplication can be used as a check for division.

104. Explain what it means when we say that division by 0 is undefined.

105. Explain the division rules for integers that are shown below using symbols.

$$\frac{+}{+} = + \qquad \frac{-}{-} = + \qquad \frac{-}{+} = - \qquad \frac{+}{-} = -$$

106. Explain the difference between *division of 0* and *division by 0*.

107. Explain why the operation of division is not commutative.

108. Is the following statement true or false: 0 divided by any number is 0. Explain your reasoning.

SECTION 1.6
Applications Introduction: Order of Operations

In mathematics, when we speak of *operations* we are referring to procedures such as addition, subtraction, multiplication, division, and squaring. In Section 1.6, you will learn important order of operations rules that must be followed when making calculations that involve more than one operation. Here are two examples that will help you see why such rules are necessary.

1. CARNIVAL TICKETS A preschool teacher has a morning session with 12 children and an afternoon session with 8 other children. She wants to give each child two tickets to the school carnival and take two tickets for herself.

a. To find the total number of tickets to request from the carnival organizer, the teacher made the calculation below. After receiving that total, passing them out to the all the children, and taking two for herself, she had 38 tickets left over. Explain why.

Two tickets for myself	plus	Two tickets for each child	times	The total number of children in both of my classes
2	+	2	·	(12 + 8)

$$= 4 \cdot (12 + 8)$$
$$= 4 \cdot (20)$$
$$= 80$$

b. How many tickets did the teacher actually need?

2. ELECTIONS Read each sentence below. According to what you read, who won the election?

a. John Hunt said, "Leslie Owens won the election."

b. "John Hunt," said Leslie Owens, "won the election."

c. Notice that both sentences contain the exact same words:

John Hunt said Leslie Owens won the election.

Why do the sentences have different meanings?

SECTION 1.6
Exponents and Order of Operations

Objectives

1. Evaluate exponential expressions.
2. Use the order of operations rules.
3. Evaluate expressions with no grouping symbols.
4. Evaluate expressions containing grouping symbols.

ARE YOU READY?

The following problems review some basic concepts that are important when working with numerical expressions.

1. Name the operations that are involved in the expression: $50 - 2(3)$

2. Name the operations that are involved in the expression: $\dfrac{2 + 6 \cdot 8}{4 + 6}$

3. Multiply: $3 \cdot 3 \cdot 3 \cdot 3$

4. Multiply: $(-5)(-5)(-5)$

Six operations can be performed with integers: addition, subtraction, multiplication, division, raising to a power, and finding a root. Quite often, we will have to **evaluate** (find the value of) expressions containing more than one operation. In that case, we need to know the order in which the operations are to be performed. That is a topic of this section.

1 Evaluate exponential expressions.

In the expression $3 \cdot 3 \cdot 3 \cdot 3 \cdot 3$, the number 3 is used as a factor 5 times. We call 3 a *repeated factor.* To express a repeated factor, we can use an **exponent.**

Exponent and Base

An **exponent** is used to indicate repeated multiplication. It tells how many times the **base** is used as a factor.

$$\underbrace{3 \cdot 3 \cdot 3 \cdot 3 \cdot 3}_{\text{Five repeated factors of 3.}} = 3^5$$

The exponent is 5.

The base is 3.

In the **exponential expression** 3^5, 3 is the base, and 5 is the exponent. The expression 3^5 is called a **power of 3.** Some examples of powers are

5^2 Read as "5 to the second power" or "5 squared."

9^3 Read as "9 to the third power" or "9 cubed."

$(-2)^5$ Read as "−2 to the fifth power."

The Language of Algebra 5^2 represents the area of a square with sides 5 units long. 4^3 represents the volume of a cube with sides 4 units long.

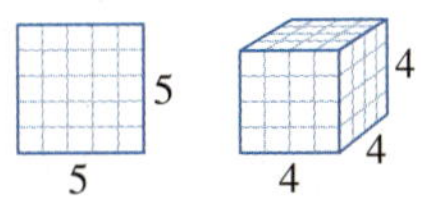

EXAMPLE 1 Write each expression using exponents: **a.** $4 \cdot 4 \cdot 4$ **b.** $(-5)(-5)(-5)(-5)(-5)$ **c.** sixteen cubed **d.** $8 \cdot 8 \cdot 15 \cdot 15 \cdot 15 \cdot 15$

Strategy We will count the number of repeated factors in each expression.

WHY An exponent can be used to represent repeated multiplication.

Solution

a. The factor 4 is repeated 3 times. We can represent this repeated multiplication with an exponential expression having a base of 4 and an exponent of 3: $4 \cdot 4 \cdot 4 = 4^3$.

b. The factor −5 is repeated 5 times: $(-5)(-5)(-5)(-5)(-5) = (-5)^5$.

c. Sixteen cubed can be written as 16^3.

d. $8 \cdot 8 \cdot 15 \cdot 15 \cdot 15 \cdot 15 = 8^2 \cdot 15^4$

Self Check 1

Write each expression using exponents:

a. $(12)(12)(12)(12)(12)(12)$

b. $2 \cdot 9 \cdot 9 \cdot 9$

c. fifty squared

d. $(-30)(-30)(-30)$

Now Try **Problem 25**

Self Check 2

Find each power:

a. 2^5
b. $(-6)^2$
c. $(-5)^3$

Now Try **Problems 29 and 33**

EXAMPLE 2 Find each power: **a.** 5^3 **b.** 10^1 **c.** $(-3)^4$ **d.** $(-3)^5$

Strategy We will identify the base to determine the repeated factor and identify the exponent to determine the number of times the factor is repeated. Then we will multiply to evaluate the expression.

WHY Exponents represent repeated multiplication.

Solution

We write the base as a factor the number of times indicated by the exponent. Then we perform the multiplication.

a. $5^3 = 5 \cdot 5 \cdot 5 = 125$ The base is 5, the exponent is 3.

b. $10^1 = 10$ The base is 10, the exponent is 1.

c. $(-3)^4 = (-3)(-3)(-3)(-3)$ Write -3 as a factor 4 times.
$= 9(-3)(-3)$ Work from left to right: $(-3)(-3) = 9$.
$= -27(-3)$ Work from left to right: $9(-3) = -27$.
$= 81$

d. $(-3)^5 = (-3)(-3)(-3)(-3)(-3)$ Write -3 as a factor 5 times.
$= 9(-3)(-3)(-3)$ Work from left to right: $(-3)(-3) = 9$.
$= -27(-3)(-3)$ Work from left to right: $9(-3) = -27$.
$= 81(-3)$ Work from left to right: $-27(-3) = 81$.
$= -243$

Caution! Don't make the mistake of multiplying the base and the exponent.

Incorrect	Correct
~~$5^3 = 5 \cdot 3$~~	$5^3 = 5 \cdot 5 \cdot 5$
	$= 125$

Using Your CALCULATOR **Finding a Power**

On a scientific calculator, we can use the squaring key $\boxed{x^2}$ to find the square of a number, and we can use the exponential key $\boxed{y^x}$ (on some calculators labeled x^y) to raise a number to a power. For example, to evaluate 125^2 and 2^{10} using a scientific calculator, we enter these numbers and press these keys.

125 $\boxed{x^2}$ → 15625

2 $\boxed{y^x}$ 10 $\boxed{=}$ → 1024

Using a graphing or direct-entry calculator, we can evaluate 125^2 and 2^{10} by pressing these keys.

125 $\boxed{x^2}$ $\boxed{\text{ENTER}}$ → 125² 15625

2 $\boxed{\wedge}$ 10 $\boxed{\text{ENTER}}$ → 2^10 1024

We have found that $125^2 = 15{,}625$ and 2^{10} 12 $= 1{,}024$.

We can now make some observations about raising a negative number to an *even power* (2, 4, 6, 8, and so on) and raising a negative number to an *odd power* (1, 3, 5, 7, and so on). In part c of Example 2, we raised -3 to an even power, and the result was positive. In part d, we raised -3 to an odd power, and the result was negative. These results illustrate the following general rule.

Even and Odd Powers of a Negative Number

When a negative number is raised to an even power, the result is positive.

When a negative number is raised to an odd power, the result is negative.

Although the expressions $(-4)^2$ and -4^2 look alike, they are not. When we find the value of each expression, it becomes clear that they are not equivalent. We read $(-4)^2$ as "negative four squared" and -4^2 as "the opposite of the square of four."

$(-4)^2 = (-4)(-4)$ The base is The base is 4, the exponent is 2.

$= 16$

$-4^2 = -(4 \cdot 4)$ The base is 4, the exponent is 2.

$= -16$

Different results

Any real number can be used as a base. However, the base of an exponential expression *does not include* the negative sign unless parentheses are used.

-7^3 Positive base: 7

$(-7)^3$ Negative base: -7

EXAMPLE 3 Evaluate: -2^4

Strategy We will rewrite the expression as a product of repeated factors and then perform the multiplication. We must be careful when identifying the base. It is 2, not -2.

WHY Since there are no parentheses around -2, the base is 2.

Solution

$-2^4 = -(2 \cdot 2 \cdot 2 \cdot 2)$ Read as "the opposite of the fourth power of two."

$= -16$ Do the multiplication within the parentheses to get 16. Then write the opposite of that result.

Self Check 3

Evaluate: -5^4

Now Try **Problem 39**

Using Your CALCULATOR **Raising a Negative Number to a Power**

We can find powers of negative integers, such as $(-5)^6$, using a calculator. The keystrokes that are used to evaluate such expressions vary from model to model, as shown below. You will need to determine which keystrokes produce the positive result that we would expect when raising a negative number to an even power.

5 [+/−] [y^x] 6 [=] Some calculators don't require the parentheses to be entered.

[(] 5 [+/−] [)] [y^x] 6 [=] Other calculators require the parentheses to be entered.

[(] [(−)] 5 [)] [^] 6 [ENTER] 15625

From the calculator display, we see that $(-5)^6 = 15{,}625$.

2 Use the order of operations rules.

Suppose you have been asked to contact a friend if you see a Rolex watch for sale when you are traveling in Europe. While in Switzerland, you find the watch and send the text message shown on the left. The next day, you get the response shown on the right.

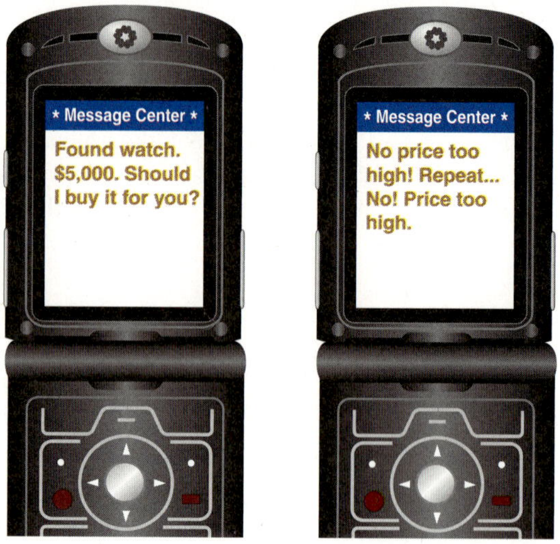

(a) (b)

Something is wrong. The first part of the response (No price too high!) says to buy the watch at any price. The second part (No! Price too high.) says not to buy it, because it's too expensive. The placement of the exclamation point makes us read the two parts of the response differently, resulting in different meanings. When reading a mathematical statement, the same kind of confusion is possible. For example, consider the expression

$$2 + 3 \cdot 6$$

which contains two operations: addition and multiplication. We can consider doing the calculations in two ways. We can add first and then multiply. Or we can multiply first and then add. However, we get different results.

Method 1: Add first		**Method 2: Multiply first**	
$2 + 3 \cdot 6 = 5 \cdot 6$	Add 2 and 3 first.	$2 + 3 \cdot 6 = 2 + 18$	Multiply 3 and 6 first.
$= 30$	Multiply 5 and 6.	$= 20$	Add 2 and 18.

If we don't establish a uniform order of operations, the expression $2 + 3 \cdot 6$ has two different values. To avoid this possibility, we always use the following set of priority rules.

Order of Operations

1. Perform all calculations within parentheses and other grouping symbols following the order listed in steps 2–4 below, working from the innermost pair to the outermost pair.
2. Evaluate all exponential expressions.
3. Perform all multiplications and divisions as they occur from left to right.
4. Perform all additions and subtractions as they occur from left to right.

When grouping symbols have been removed, repeat steps 2–4 to complete the calculation.

If a fraction is present, evaluate the expression above and the expression below the bar separately. Then do the division indicated by the fraction bar, if possible.

It isn't necessary to apply all of these steps in every problem. For example, the expression $2 + 3 \cdot 6$ does not contain any parentheses, and there are no exponential expressions. So we look for multiplications and divisions to perform. To evaluate $2 + 3 \cdot 6$ correctly, we proceed as follows:

$2 + 3 \cdot 6 = 2 + 18$ Multiply first: $3 \cdot 6 = 18$.
$= 20$ Add.

Therefore, the correct result when evaluating $2 + 3 \cdot 6$ is 20.

3 Evaluate expressions with no grouping symbols.

EXAMPLE 4 Evaluate: **a.** $3 \cdot 2^3 - 4$ **b.** $-30 - 4 \cdot 5 + 9$ **c.** $24 \div 6 \cdot 2$ **d.** $160 - 4 + 6(-2)(-3)$

Strategy We will scan the expression to determine what operations need to be performed. Then we will perform those operations, one-at-a-time, following the order of operations rules.

WHY The order of operations gives us the steps needed to find the correct result.

Solution

a. Three operations need to be performed to evaluate this expression: multiplication, raising to a power, and subtraction. By the order of operations rules, we evaluate 2^3 first.

$3 \cdot 2^3 - 4 = 3 \cdot 8 - 4$ Evaluate the exponential expression: $2^3 = 8$.
$= 24 - 4$ Do the multiplication: $3 \cdot 8 = 24$.
$= 20$ Do the subtraction.

b. This expression involves subtraction, multiplication, and addition. The order of operations rule tells us to multiply first.

$-30 - 4 \cdot 5 + 9 = -30 - 20 + 9$ Do the multiplication: $4 \cdot 5 = 20$.
$= -50 + 9$ Working from left to right, do the subtraction: $-30 - 20 = -30 + (-20) = -50$
$= -41$ Do the addition.

c. Since there are no calculations within parentheses nor are there exponents, we perform the multiplications and divisions as they occur from left to right. The division occurs before the multiplication, so it must be performed first.

$24 \div 6 \cdot 2 = 4 \cdot 2$ Working left to right, do the division: $24 \div 6 = 4$.
$= 8$ Do the multiplication.

d. Although this expression contains parentheses, there are no operations to perform within them. Since there are no exponents, we will perform the multiplications as they occur from left to right.

$160 - 4 + 6(-2)(-3) = 160 - 4 + (-12)(-3)$ Do the multiplication, working left to right: $6(-2) = -12$.
$= 160 - 4 + 36$ Complete the multiplication: $(-12)(-3) = 36$.
$= 156 + 36$ Working left to right, do the subtraction before the addition.
$= 192$ Do the addition.

Self Check 4

Evaluate:

a. $2 \cdot 3^2 + 17$
b. $-40 - 9 \cdot 4 + 10$
c. $18 \div 2 \cdot 3$
d. $240 - 8 + 3(-2)(-4)$

Now Try **Problem 41, 49, 51, and 55**

The Language of Algebra Sometimes, for problems like these, the instruction *Simplify* is used instead of *Evaluate.*

Caution! Some students think that additions are always done before subtractions. As you saw in Example 4b, this is not true. Working from left to right, we do the additions or subtractions in the order in which they occur. The same is true for multiplications and divisions as in Example 4c.

Self Check 5

Evaluate:
$240 \div (-8)(3) - 3(-2)4$

Now Try **Problem 57**

EXAMPLE 5 Evaluate: $160 \div (-4)(3) - 6(-2)3$

Strategy We will scan the expression to determine what operations need to be performed. Then we will perform those operations, one at a time, following the order of operations rules.

WHY The order of operations gives us the steps needed to find the correct result.

Solution

Although this expression contains parentheses, there are no operations to perform within them. Since there are no exponents, we perform multiplications and divisions as they occur from left to right.

$$
\begin{aligned}
160 \div (-4)(3) - 6(-2)3 &= -40(3) - 6(-2)3 && \text{Divide: } 160 \div (-4) = -40. \\
&= -120 - 6(-2)3 && \text{Multiply: } -40(3) = -120 \\
&= -120 - (-12)3 && \text{Multiply: } 6(-2) = -12. \\
&= -120 - (-36) && \text{Multiply: } (-12)3 = -36. \\
&= -120 + 36 && \text{Write the subtraction as addition of the opposite.} \\
&= -84 && \text{Add.}
\end{aligned}
$$

Caution! A common mistake in Example 5 is to forget to work from left to right and incorrectly perform the multiplication before the division.

4 Evaluate expressions containing grouping symbols.

Grouping symbols are mathematical punctuation marks. They help determine the order in which an expression is to be evaluated. Examples of grouping symbols are parentheses ( ), brackets [ ], absolute value symbols | |, and the fraction bar —.

Self Check 6

Evaluate: $(12 - 6)^3$

Now Try **Problem 59**

EXAMPLE 6 Evaluate: $(6 - 3)^2$

Strategy We will perform the operation(s) within the parentheses first. When there is more than one operation to perform within the parentheses, we follow the order of operations rules.

WHY This is the first step of the order of operations.

Solution

This expression contains parentheses. By the rules for the order of operations, we must perform the operation within the parentheses first.

$(6 - 3)^2 = 3^2$ Subtract within the parentheses: $6 - 3 = 3$.

$= 9$ Evaluate the exponential expression.

EXAMPLE 7 Evaluate: $5^3 + 2(-8 - 3 \cdot 2)$

Strategy We will perform the operation(s) within the parentheses first. When there is more than one operation to perform within the parentheses, we follow the order of operations rules.

WHY This is the first step of the order of operations.

Solution

First, we perform the operations within the parentheses in the proper order.

$5^3 + 2(-8 - 3 \cdot 2) = 5^3 + 2(-8 - 6)$ Multiply within the parentheses: $3 \cdot 2 = 6$.

$= 5^3 + 2(-14)$ Subtract within the parentheses: $-8 - 6 = -8 + (-6) = -14$.

$= 125 + 2(-14)$ Evaluate the exponential expression: $5^3 = 125$.

$= 125 + (-28)$ Multiply: $2(-14) = -28$.

$= 97$ Add.

Self Check 7

Evaluate: $1^3 + 6(-6 - 3 \cdot 0)$

Now Try **Problem 63**

Success Tip Multiplication is indicated when a number is next to a parentheses, bracket, or absolute value symbols.

Expressions can contain two or more pairs of grouping symbols. To evaluate the following expression, we begin by working within the innermost pair of grouping symbols. Then we work within the outermost pair.

Innermost pair

$-4[-2 - 3(4 - 8^2)] - 2$

Outermost pair

The Language of Algebra When one pair of grouping symbols is inside another pair, we say that those grouping symbols are *nested,* or *embedded.*

EXAMPLE 8 Evaluate: $-4[-2 - 3(4 - 8^2)] - 2$

Strategy We will work within the parentheses first and then within the brackets. At each stage, we follow the order of operations rules.

WHY By the order of operations, we must work from the *innermost* pair of grouping symbols to the *outermost.*

Self Check 8

Evaluate:
$-5[2(5^2 - 15) + 4] - 10$

Now Try **Problem 74**

Solution

We work within the innermost grouping symbols (the parentheses) first.

$-4[-2 - 3(4 - 8^2)] - 2$

$= -4[-2 - 3(4 - 64)] - 2$ — Evaluate the exponential expression within the parentheses: $8^2 = 64$.

$= -4[-2 - 3(-60)] - 2$ — Subtract within the parentheses: $4 - 64 = 4 + (-64) = -60$.

$= -4[-2 - (-180)] - 2$ — Multiply within the brackets: $3(-60) = -180$.

$= -4(178) - 2$ — Subtract within the brackets: $-2 - (-180) = -2 + 180 = 178$.

$= -712 - 2$ — Multiply.

$= -714$ — Subtract: $-712 - 2 = -712 + (-2) = -714$.

Self Check 9

Evaluate: $\dfrac{-4(-2 + 8) + 6}{8 - 5(-2)}$

Now Try **Problem 75**

EXAMPLE 9

Evaluate: $\dfrac{-3(3 + 2) + (-4)}{7 - 3(-4)}$

Strategy We will evaluate the expression above and the expression below the fraction bar separately. Then we will simplify the fraction, if possible.

WHY Fraction bars are grouping symbols. They group the numerator and denominator. The expression could be written as $[-3(3 + 2) + (-4)] \div [7 - 3(-4)]$.

Solution

We simplify the numerator and the denominator separately.

$\dfrac{-3(3 + 2) + (-4)}{7 - 3(-4)} = \dfrac{-3(5) + (-4)}{7 - (-12)}$ — In the numerator, add within the parentheses. In the denominator, multiply.

$= \dfrac{-15 + (-4)}{7 + 12}$ — In the numerator, multiply. In the denominator, write the subtraction as addition of the opposite of −12, which is 12.

$= \dfrac{-19}{19}$ — Perform the additions.

$= -1$

Success Tip The order of operations are built in to most calculators. A left parenthesis key (and a right parenthesis key) should be used when grouping symbols, including a fraction bar, are in the problem.

Self Check 10

Evaluate: $10^3 + 3|24 - 25|$

Now Try **Problem 86**

EXAMPLE 10

Evaluate: $10|9 - 15| - 2^5$

Strategy The absolute value bars are grouping symbols. We will perform the calculation within them first.

WHY By the order of operations, we must perform all calculations within parentheses and other grouping symbols (such as absolute value bars) first.

Solution

Since the absolute value bars are grouping symbols, we perform the calculation within them first.

$$10|9 - 15| - 2^5 = 10|-6| - 2^5$$ Subtract: $9 - 15 = 9 + (-15) = -6$.

$$= 10(6) - 2^5$$ $10|-6|$ means 10 times $|-6|$. Find the absolute value: $|-6| = 6$.

$$= 10(6) - 32$$ Evaluate the exponential expression: $2^5 = 32$.

$$= 60 - 32$$ Multiply.

$$= 28$$

ANSWERS TO SELF CHECKS

1. a. 12^6 **b.** $2 \cdot 9^3$ **c.** 50^2 **d.** $(-30)^3$ **2. a.** 32 **b.** 36 **c.** -125 **3.** -625
4. a. 35 **b.** -66 **c.** 27 **d.** 256 **5.** -66 **6.** 216 **7.** -35 **8.** -130
9. -1 **10.** 1,003

SECTION 1.6 STUDY SET

VOCABULARY

Fill in the blanks.

1. In the exponential expression 3^2, the ______, is 3, and 2 is the ________.

2. 10^2 can be read as ten ________, and 10^3 can be read as ten ______.

3. 7^5 is the fifth ______ of seven.

4. An ________ is used to represent repeated multiplication.

5. The rules for the ______ of operations guarantee that an evaluation of a numerical expression will result in a single answer.

6. In the expression $9 + 6[22 - (6 - 1)]$, the ________ are the innermost grouping symbols, and the brackets are the _______ grouping symbols.

CONCEPTS

7. Given: $4 + 5 \cdot 6$

a. What operations does this expression contain?

b. Evaluate the expression in two different ways, and state the two possible results.

c. Which result from part b is correct, and why?

8. a. What repeated multiplication does 5^3 represent?

b. How can we represent the repeated addition $3 + 3 + 3 + 3 + 3$ in a simpler form?

9. What operation is indicated?

$$2 + 9\overset{\downarrow}{|}5 - (2 + 4)|$$

10. In the expression $-8 + 2[15 - (-6 + 1)]$, which grouping symbols are innermost and which are outermost?

11. a. What operations does the expression $12 + 5^2(-3)$ contain?

b. In what order should they be performed?

12. a. What operations does the expression $20 - (-2)^2 + 3(-1)$ contain?

b. In what order should they be performed?

13. Consider the expression $\frac{36 - 4(7)}{2(10 - 8)}$. In the numerator, what operation should be done first? In the denominator, what operation should be done first?

14. Explain the differences in evaluating $4 \cdot 2^2$ and $(4 \cdot 2)^2$.

15. To evaluate each expression, what operation should be performed first?

a. $-80 - 3 + 5 - 2^2$

b. $-80 - (3 + 5) - 2^2$

c. $-80 + 3 + (5 - 2)^2$

d. $-80 \div 4 \cdot 2$

e. $-80 - 4 + 2$

16. To evaluate each expression, what operation should be performed first?

a. $(65 - 3)^3$

b. $65 - 3^3$

c. $6(5) - (3)^3$

NOTATION

17. Write an exponential expression with a base of 12 and an exponent of 6.

18. Give the name of each grouping symbol: ( ), [ ], | |, and —.

19. a. In the expression $(-5)^2$, what is the base?

b. In the expression -5^2, what is the base?

20. Write each expression using symbols. Then evaluate it.

a. Negative two squared

b. The opposite of the square of two

Complete each evaluation.

21. $50 + 6 \cdot 3^2 = 50 + 6 \cdot \square$
$= 50 + \square$
$= 104$

22. $-100 - (25 - 8 \cdot 2) = -100 - (25 - \square)$
$= -100 - \square$
$= -109$

23. $-19 - 2[(1 + 2) \cdot 3] = -19 - 2[\square \cdot 3]$
$= -19 - 2(\square)$
$= -19 - \square$
$= -37$

24. $\dfrac{46 - 2^3}{-3(5) - 4} = \dfrac{46 - \square}{\square - 4}$
$= \dfrac{\square}{\square}$
$= -2$

GUIDED PRACTICE

Write each product using exponents. See Example 1.

25. $3 \cdot 3 \cdot 3 \cdot 3$

26. $(-7)(-7)(-7)(-7)(-7)(-7)$

27. $10 \cdot 10 \cdot 12 \cdot 12 \cdot 12$

28. $5(5)(5)(11)(11)$

Find each power. See Example 2.

29. 7^2 **30.** 11^3

31. 6^3 **32.** 6^4

33. 5^4 **34.** $(-5)^3$

Find each power. See Example 3.

35. $(-2)^3$ **36.** $(-4)^4$

37. -3^3 **38.** -5^3

39. -6^2 **40.** -4^4

Evaluate each expression. See Examples 4 and 5.

41. $3 - 5 \cdot 4$ **42.** $-4 \cdot 6 + 5$

43. $3 \cdot 8^2 - 5$ **44.** $3 \cdot 4^2 - 8$

45. $32 - 16 \div 4 + 2$ **46.** $60 - 20 \div 10 + 5$

Evaluate each expression. See Examples 6–7.

47. $3^3 - 2^3$ **48.** $9 \cdot 5 - 6 \div 3$

49. $3^2 \cdot 5 - 6 \div 3$ **50.** $2^3 \cdot 5 - 4 \div 2$

51. $12 \div 3 \cdot 2$ **52.** $18 \div 6 \cdot 3$

53. $-22 - 15 + 3$ **54.** $-33 - 8 + 10$

55. $-2(9) - 2(5)(10)$ **56.** $-6(7) - 3(-4)(-2)$

57. $2 \cdot 3^2 + 5 \cdot 2^3$ **58.** $4 \cdot 2^5 - 3 \cdot 5^2$

Evaluate each expression. See Example 6.

59. $(-5 - 2)^2$ **60.** $(-3 - 5)^2$

61. $(12 - 2)^3$ **62.** $(10 - 3)^2$

Evaluate each expression. See Example 7.

63. $-4(6 + 5)$ **64.** $-3(5 - 4)$

65. $200 - (-6 + 5)^3$ **66.** $19 - (-45 + 41)^3$

67. $-6(130 - 4^3)$ **68.** $-5(150 - 3^3)$

69. $5 \cdot 2^2 \cdot 4 - 30$ **70.** $2 + (3 \cdot 2^2 \cdot 4)$

Evaluate each expression. See Example 8.

71. $-3[5^2 - (7 - 3)^2]$ **72.** $3 - [3^3 + (3 - 1)^3]$

73. $5 + (4^2 - 2^3)^2$ **74.** $(-5)^3[4(2^3 - 3^2)]^2$

Evaluate each expression. See Example 9.

75. $\dfrac{5 \cdot 50 - 160}{-9}$ **76.** $\dfrac{5(68 - 32)}{-9}$

77. $\dfrac{(4^3 - 10) + (-4)}{5^2 - (-4)(-5)}$ **78.** $\dfrac{(6 - 5)^4 - (-21)}{(-9)(-3) - 4^2}$

79. $\dfrac{72 - (2 - 2 \cdot 1)}{10^2 - (90 + 2^2)}$ **80.** $\dfrac{13^2 - 5^2}{-3(5 - 9)}$

81. $\dfrac{40 \div 2 - 5 \cdot 2}{3^2 - (-1)}$ **82.** $\dfrac{(5 - 2)^2 - (2 - (-1))}{5 \cdot 2 + (-7)}$

Evaluate each expression. See Example 10.

83. $-2|4 - 8|$

84. $-5|1 - 8|$

85. $|7 - 8(4 - 7)|$

86. $|9 - 5(1 - 8)|$

87. $\dfrac{|6 - 4| + 2|-4|}{26 - 2^4}$

88. $\dfrac{4|9 - 7| + |-7|}{3^2 - 2^2}$

89. $\dfrac{(3 + 5)^2 + |-2|}{-2(5 - 8)}$

90. $\dfrac{|-25| - 8(-5)}{2^4 - 29}$

TRY IT YOURSELF

Evaluate each expression.

91. $-(-6)^4$

92. $-(-7)^2$

93. $-4(6 + 5)$

94. $-3(5 - 4)$

95. $4^2 - (-2)^2$

96. $3 + (-5)^2$

97. $12 + 2\left(-\dfrac{9}{3}\right) - (-2)$

98. $2 + 3\left(-\dfrac{25}{5}\right) - (-4)$

99. $1(2)(3)(-4)$

100. $3(4)(5)(-6)$

101. $[6(5) - 5(5)]4$

102. $5[9(2) - 2(8)]$

103. $(17 - 5 \cdot 2)^3$

104. $(4 + 2 \cdot 3)^4$

105. $-5(-2)^3(3)^2$

106. $-3(-2)^5(2)^2$

107. $-2\left(\dfrac{15}{-5}\right) - \dfrac{6}{2} + 9$

108. $-6\left(\dfrac{25}{-5}\right) - \dfrac{36}{9} + 1$

109. $5(10 + 2) - 1$

110. $14 + 3(7 - 5)$

111. $64 - 6[15 + (-3)3]$

112. $4 + 2[26 + 5(-3)]$

113. $(-2)^3\left(\dfrac{-6}{2}\right)(-1)$

114. $(-3)^3\left(\dfrac{-4}{2}\right)(-1)$

115. $\dfrac{-7 - 3^2}{2 \cdot 4}$

116. $\dfrac{-5 - 3^3}{2^3}$

117. $\dfrac{18 - [2 + (1 - 6)]}{16 - (-4)^2}$

118. $\dfrac{6 - [6(-1) - 88]}{4 - 2^2}$

119. $3 + 2[-1 - 4(5)]$

120. $4 + 2[-7 - 3(9)]$

121. $-(2 \cdot 3 - 4)^3$

122. $-(3 \cdot 5 - 2 \cdot 6)^2$

123. $\dfrac{-5^2 \cdot 10 + 5 \cdot 2^5}{-5 - 3 - 1}$

124. $\dfrac{(-6^2 - 2^4 \cdot 2) + 5}{-4 - 3}$

125. $-\left(\dfrac{40 - 1^3 - 2^4}{3(2 + 5) + 2}\right)$

126. $-\left(\dfrac{8^2 - 10}{2(3)(4) - 5(3)}\right)$

127. $\dfrac{3(3{,}246 - 1{,}111)}{561 - 546}$

128. $54^3 - 16^4 + 19(3)$

LOOK ALIKES . . .

129. a. $(-7 - 4)(-2)$ **b.** $(-7 - 4) - 2$

130. a. $2 \cdot 3^3$ **b.** $(2 \cdot 3)^3$

131. a. $-100 \div 5 \cdot 2$ **b.** $-100 \div (5 \cdot 2)$

132. a. $8 + 3[-2 - (6 + 1)]$

b. $(8 + 3)[-2 - (6 + 1)]$

APPLICATIONS

133. CHAIN LETTERS A store owner sent two friends a letter advertising her store's low prices. The ad closed with the following request: "Please send a copy of this letter to two of your friends."

a. Assume that all those receiving letters respond and that everyone in the chain receives just one letter. Complete the table.

b. How many letters will be circulated in the tenth level of the mailing?

Level	Numbers of letters circulated
1st	$2 = 2^1$
2nd	= 2
3rd	= 2
4th	= 2

134. SHOPPING John bought three neckties for \$8 each and two shirts for \$28 each. The sales tax on the purchase was \$4. He gave the cashier a \$100 bill. Complete the following expression to determine the amount of change he should receive. Then evaluate it.

$\square - (3 \cdot \square + 2 \cdot \square + \square)$

135. SCRABBLE Illustration (a) on the next page shows a portion of the game board before and illustration (b) shows it after the word *QUARTZY* is played. Complete the expression below to determine the number of points received for playing that word. Then evaluate the expression. (The number on each tile gives the point value of that letter.)

$\square[10 + \square + \square + \square + \square + 2(\square) + \square]$

(a)

(b)

136. WRAPPING GIFTS Complete the expression below to determine how many inches of ribbon are needed to wrap the package. Then evaluate the expression. (It takes 15 inches of ribbon to make the bow.)

$$(2 \cdot \square + 2 \cdot \square + 4 \cdot \square) + \square$$

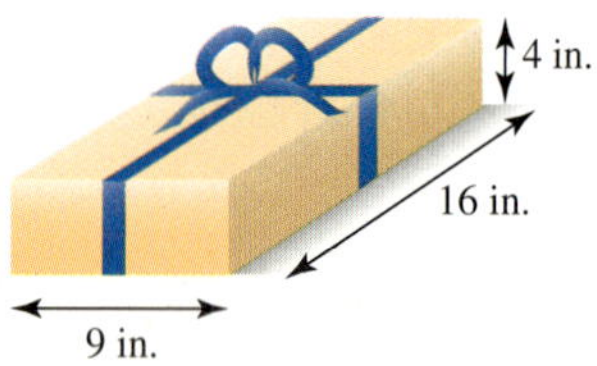

137. SPREADSHEETS This spreadsheet contains data collected by a chemist. For each row, the sum of the values in columns A and B is to be subtracted from the product of 6 and the value in column C. That result is then to be divided by 12 and entered in column D. Use this information to complete the spreadsheet.

	A	B	C	D
1	20	4	8	
2	9	3	16	
3	1	5	11	

138. DOG SHOWS The final score for each dog competing in a toy breeds competition is computed by dividing the sum of the judges' marks, after the highest and lowest have been dropped, by 6. See the table.

a. What was their order of finish?

b. Did any judge rate all the dogs the same?

Judge	1	2	3	4	5	6	7	8
Terrier	14	11	11	10	12	12	13	13
Pekingese	10	9	8	11	11	12	9	10
Pomeranian	15	14	13	11	14	12	10	14

WRITING

139. Explain the difference between 2^3 and 3^2.

140. Explain why rules for the order of operations are necessary.

141. What does it mean when we say perform all additions and subtractions *as they occur from left to right?*

142. In what settings do you encounter or use the concept of arithmetic mean (average) in your everyday life?

143. Explain the error. What is the correct answer?

$$\begin{aligned} 40 \div 4 \cdot 2 &= 0 \div 8 \\ &= 5 \end{aligned}$$

144. Explain the error. What is the correct answer?

$$\begin{aligned} 5 + 3(2 - 6) &= 5 + 3(-4) \\ &= 8(-4) \\ &= -32 \end{aligned}$$

SECTION 1.7

Algebraic Expressions

Objectives

1. Use the basic vocabulary and notation of algebra.
2. Translate word phrases to algebraic expressions.
3. Write algebraic expressions to represent unknown quantities.

ARE YOU READY?

The following problems review some basic concepts that are important when working with variables and algebraic expressions.

What operation (addition, subtraction, multiplication, or division) is associated with each word?

1. product
2. difference
3. quotient
4. sum

Consider a bag of M&M's. What operation describes each action?

5. removing 3 pieces
6. doubling the number of pieces
7. increasing the number of pieces by 12
8. Sharing equally among 4 people

With an in-depth study of the integers now completed, it's time to turn our attention to algebra. *Algebra* is the language of mathematics. It has its own vocabulary (words) and notation (symbols). In this section, you will learn how to think and write in this language, using its most basic component - a *variable.*

1 Use the basic vocabulary and notation of algebra.

In algebra, we often use **tables** to show relationships between quantities. For example, the table below lists the number of calories a 160-pound adult burns during 10, 20, 30, and 40 minutes of snowboarding. For a workout of, say, 30 minutes, we locate 30 in the left column and then scan across the table to see that 300 calories are burned.

©Ipatov/Shutterstock.com

Minutes snowboarding	Calories burned
10	100
20	200
30	300
40	400

From the table, we see that there is a relationship between the number of calories burned and the number of minutes snowboarding. Using words, we can express this relationship as a **verbal model:**

> "The number of calories burned is ten times the number of minutes snowboarding."
>
> Since the word **product** indicates the result of a multiplication, we can also write:
>
> "The number of calories burned is the *product* of ten and the number of minutes snowboarding."

Many symbols used in arithmetic are also used in algebra. For example, a $+$ symbol is used to indicate addition, a $-$ symbol is used to indicate subtraction, and an $=$ symbol means *is equal to.*

Since the letter x is often used in algebra and could be confused with the multiplication symbol $\times$, we usually write multiplication using a **raised dot** or **parentheses.**

Symbols for Multiplication

$\times$	Times symbol	$6 \times 4 = 24$
$\cdot$	Raised dot	$6 \cdot 4 = 24$
( )	Parentheses	$(6)4 = 24$ or $6(4) = 24$ or $(6)(4) = 24$

In algebra, the symbol most often used to indicate division is the **fraction bar.**

Symbols for Division

$\div$	Division symbol	$24 \div 4 = 6$
$\overline{)\ }$	Long division	$4\overset{6}{\overline{)24}}$
—	Fraction bar	$\frac{24}{4} = 6$

The Language of Algebra

The collection of symbols and write-up forms used in this course is called the **notation** of algebra.

Another way to describe the relationship between calories burned and snowboarding time uses *variables.* A variable is a letter (or symbol) that stands for a number. If we let the letter m represent the number of minutes snowboarding, then the number of calories burned is ten times m, written $10m$. In this notation, the number 10 is an example of a **constant** because it does not change value.

The Language of Algebra

Since the number of minutes snowboarding can *vary,* or change, it is represented using a **variable**.

When multiplying a variable by a number, or a variable by another variable, we can omit the symbol for multiplication. For example,

$10m$ means $10 \cdot m$ $\quad$ xy means $x \cdot y$ $\quad$ $8abc$ means $8 \cdot a \cdot b \cdot c$

We call $10m$, xy, and $8abc$ *algebraic expressions.*

Algebraic Expressions

Variables and/or numbers can be combined with the operations of addition, subtraction, multiplication, and division to create **algebraic expressions.**

The Language of Mathematics We often refer to *algebraic expressions* as simply *expressions.*

Here are some examples of algebraic expressions.

$4a + 7$ — This expression is a combination of the numbers 4 and 7, the variable a, and the operations of multiplication and addition.

$\frac{10 - y}{3}$ — This expression is a combination of the numbers 10 and 3, the variable y, and the operations of subtraction and division.

Algebraic expressions can contain two (or more) variables.

$15mn(2m)$ — This expression is a combination of the numbers 15 and 2, the variables m and n, and the operation of multiplication.

In the snowboarding example, if we let the letter c stand for the number of calories burned, we can translate the verbal model to mathematical symbols.

The number of calories burned	is	ten	times	the number of minutes snowboarding.
c	$=$	10	$\cdot$	m

The statement $c = 10 \cdot m$, or more simply, $c = 10m$, is called an *equation.* An **equation** is a mathematical sentence that contains an $=$ symbol. The $=$ symbol indicates that the expressions on either side of it have the same value. Other examples of equations are

$3 + 5 = 8 \quad x + 5 = 20 \quad 17 - 2r = 14 + 3r \quad P = 100 - d$

The Language of Algebra

The equal symbol $=$ can be represented by verbs such as:

is *are* *gives* *yields*

The symbol $\neq$ is read as *"is not equal to."*

Caution! Throughout this course you will be working with *expressions* and *equations.* It is important to know the difference between them. An equation contains an $=$ symbol. An expression does not.

2 Translate word phrases to algebraic expressions.

In order to solve application problems, which are almost always given in words, we must translate those words into mathematical symbols. The following tables show how key words and phrases can be translated into algebraic expressions.

Addition	
the sum of a and 8	$a + 8$
4 plus c	$4 + c$
16 added to m	$m + 16$
4 more than t	$t + 4$
20 greater than F	$F + 20$
T increased by r	$T + r$
exceeds y by 35	$y + 35$

Subtraction	
the difference of 23 and P	$23 - P$
550 minus h	$550 - h$
18 less than w	$w - 18$
7 decreased by j	$7 - j$
M reduced by x	$M - x$
12 subtracted from L	$L - 12$
5 less f	$5 - f$

Caution! Be careful when translating subtraction. Order is important. For example, when a translation involves the phrase *less than,* note how the terms are reversed.

18 less than w

$w - 18$

Multiplication	
the product of 4 and x	$4x$
20 times B	$20B$
twice r	$2r$
double the amount a	$2a$
triple the profit P	$3P$
three-fourths of m*	$\frac{3}{4}m$

* This translation is discussed in more detail in Module DMA 020.

Division	
the quotient of R and 19	$\frac{R}{19}$
s divided by d	$\frac{s}{d}$
k split into 4 equal parts	$\frac{k}{4}$
the ratio of c to d*	$\frac{c}{d}$

* This translation is discussed in more detail in Module DMA 030.

Caution! Be careful when translating division. As with subtraction, order is important. For example, s divided by d is *not* written $\frac{d}{s}$.

Self Check 1

Write each phrase as an algebraic expression:

a. 80 less than the total t

b. 6 more than T

c. the difference of twice a and 15, squared

Now Try **Problems 23, 24, and 25**

EXAMPLE 1 Write each phrase as an algebraic expression:

a. twice the profit P

b. 5 less than the capacity c

c. the product of the weight w and 2,000, increased by 300

Strategy We will begin by identifying any key words or phrases.

WHY Key words or phrases can be translated to mathematical symbols.

Solution

a. Key word: twice **Translation:** multiplication by 2

The algebraic expression is: $2P$.

b. Key phrase: *less than* **Translation:** subtraction

Sometimes thinking in terms of specific numbers makes translating easier. Suppose the capacity was 100. Then 5 *less than* 100 would be $100 - 5$. If the capacity is c, then we need to make c, 5 less. The algebraic expression is: $c - 5$.

Caution! $5 < c$ is the translation of the statement 5 *is less than* the capacity c and not 5 *less than* the capacity c.

c. Key phrase: *product of* **Translation:** multiplication

Key phrase: *increased by* **Translation:** addition

In the given wording, the comma after 2,000 means w is first multiplied by 2,000; then 300 is added to that product. The algebraic expression is: $2{,}000w + 300$.

3 Write algebraic expressions to represent unknown quantities.

To solve application problems, we let a variable stand for an unknown quantity. We can use the translation skills just discussed to describe any other unknown quantities in the problem by using algebraic expressions.

EXAMPLE 2 ***Banking*** Javier deposited d dollars in his checking account. He deposited \$500 more than that in his savings account. Write an algebraic expression that represents the amount that he deposited in the savings account.

Strategy We will carefully read the problem, looking for a key word or key phrase.

WHY Then we can translate the key word (or phrase) to mathematical symbols to represent the unknown amount that Javier deposited in the savings account.

Solution The deposit that Javier made to the savings account was \$500 *more than* the d dollars he deposited in his checking account.

Key phrase: *more than* **Translation:** add

The number of dollars he deposited in the savings account was $d + 500$.

Self Check 2

COMMUTING It takes Val m minutes to get to work if she drives her car. If she takes the bus, her travel time exceeds this by 15 minutes. Write an algebraic expression that represents the time (in minutes) that it takes her to get to work by bus.

Now Try **Problem 51**

When solving application problems, we are rarely told which variable to use. We must decide what the unknown quantities are and how to represent them using variables.

EXAMPLE 3 ***Sports Memorabilia*** The value of the baseball card shown on the right is 4 times that of the football card. Choose a variable to represent the value of one card. Then write an algebraic expression that represents the value of the other card.

Strategy There are two unknowns—the value of the baseball card and the value of the football card. We will let v = the value of the football card.

WHY The words of the problem tell us that the value of the baseball card is related to (based on) the value of the football card.

Solution The baseball card's value is *4 times* that of the football card.

Key phrase: 4 *times* **Translation:** multiply by 4

Therefore, $4v$ is the value of the baseball card.

Self Check 3

CLOTHING SALES The sale price of a sweater is \$20 less than the regular price. Choose a variable to represent one price. Then write an algebraic expression that represents the other price.

Now Try **Problem 63**

Caution! A variable is used to represent an unknown number. Therefore, in the previous example, it would be incorrect to write, "Let v = football card," because the football card is not a number. We need to write, "Let v = the *value* of the football card."

Self Check 4

LOTTOS The payoff for a winning lottery ticket is to be split equally among fifteen friends. Write an algebraic expression that represents each person's share of the prize (in dollars).

Now Try **Problem 65**

EXAMPLE 4 ***Swimming*** A pool is to be sectioned into eight equally wide swimming lanes. Write an algebraic expression that represents the width of each lane.

Strategy There are two unknowns—the width of the pool and the width of each lane. We will begin by letting w = the width of the pool (in feet), as shown in the illustration.

WHY The width of each lane is related to (based on) the width of the pool.

Solution The width of the pool is sectioned into *eight equally wide lanes.*

Key phrase: *eight equally wide lanes* **Translation:** division by 8

Therefore, the width of each lane is $\frac{w}{8}$ feet.

Self Check 5

ELECTIONS In an election, the incumbent received 55 fewer votes than three times the challenger's votes. Choose a variable to represent the number of votes received by one candidate. Then write an algebraic expression that represents the number of votes received by the other.

Now Try **Problem 67**

EXAMPLE 5 ***Enrollments*** Second semester enrollment in a nursing program was 32 more than twice that of the first semester. Choose a variable to represent the enrollment for one of the semesters. Then write an algebraic expression that represents the enrollment for the other semester.

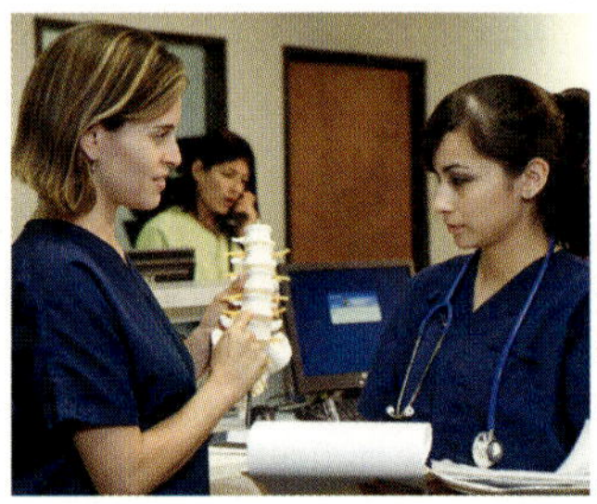
Somos/Veer/Getty Images

Strategy There are two unknowns—the enrollment for the first semester and the enrollment for the second semester. We will begin by letting x = the enrollment for the first semester.

WHY The second-semester enrollment is related to (based on) the first-semester enrollment.

Solution

Key phrase: *more than* **Translation:** addition

Key phrase: *twice that* **Translation:** multiplication by 2

The second semester enrollment was $2x + 32$.

Self Check 6

SCHOLARSHIPS Part of a $900 donation to a college went to the scholarship fund, the rest to the building fund. Choose a variable to represent the amount donated to one of the funds. Then write an expression that represents the amount donated to the other fund.

Now Try **Problem 71**

EXAMPLE 6 ***Painting*** A 10-inch-long paintbrush has two parts: a handle and bristles. Choose a variable to represent the length of one of the parts. Then write an algebraic expression to represent the length of the other part.

Strategy There are two approaches. We can let h = the length of the handle or we can let b = the length of the bristles.

WHY Both the length of the handle and the length of the bristles are unknown.

Solution Refer to the first drawing on the right. If we let h = the length of the handle (in inches), then the length of the bristles is $10 - h$.

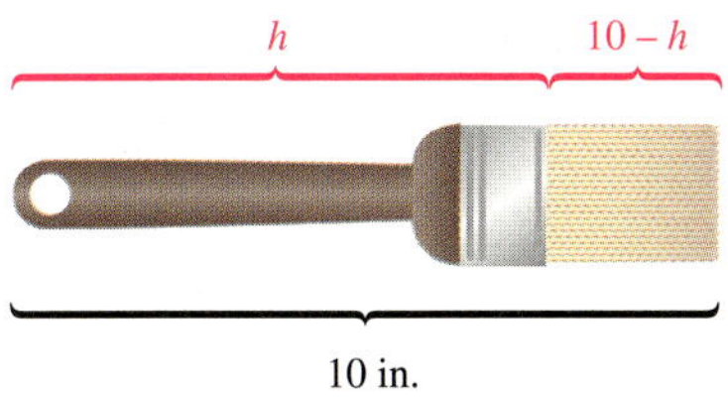

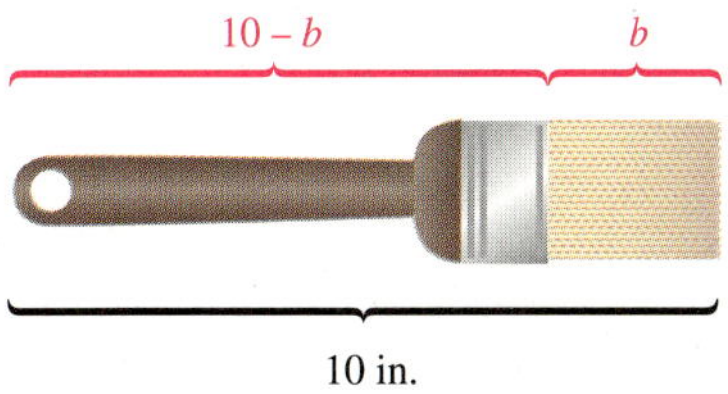

Now refer to the second drawing. If we let b = the length of the bristles (in inches), then the length of the handle is $10 - b$.

Sometimes we must analyze the wording of a problem carefully to detect hidden operations.

EXAMPLE 7 ***Engineering*** The Golden Gate Bridge was completed 28 years before the Houston Astrodome was opened. The CN Tower in Toronto was built 10 years after the Astrodome. Write algebraic expressions to represent the ages (in years) of each of these engineering wonders. (Source: Wikipedia)

Strategy There are three unknowns—the ages of the Golden Gate Bridge, the Astrodome, and the CN tower. We will begin by letting x = the age of the Astrodome (in years).

WHY The ages of the Golden Gate Bridge and the CN Tower are both related to (based on) the age of the Astrodome.

Solution Reading the problem carefully, we find that the Golden Gate Bridge was built 28 years before the dome, so its age is more than that of the Astrodome.

Key phrase: *more than* **Translation:** add

In years, the age of the Golden Gate Bridge is $x + 28$.

The CN Tower was built 10 years after the dome, so its age is less than that of the Astrodome.

Key phrase: *less than* **Translation:** subtract

In years, the age of the CN Tower is $x - 10$.

The results are summarized in the table at the right.

Engineering feat	Age
Astrodome	x
Golden Gate Bridge	$x + 28$
CN Tower	$x - 10$

Self Check 7

FAMOUS BILLS Bill Cosby was born 9 years before Bill Clinton. Bill Gates was born 9 years after Bill Clinton. Write algebraic expressions to represent the ages of each of these famous men. (Source: celebritybirthdaylist.com)

Now Try **Problem 75**

EXAMPLE 8 ***Packaging*** Write an algebraic expression that represents the number of eggs in d dozen.

Strategy First, we will determine how many eggs are in 1 dozen, 2 dozen, and 3 dozen.

WHY There are no key words or phrases in the problem. It will be helpful to consider some specific cases to determine which operation (addition, subtraction, multiplication, or division) is called for.

Solution If we calculate the number of eggs in 1 dozen, 2 dozen, and 3 dozen (as shown in the table on the next page), a pattern becomes apparent.

Number of dozen	Number of eggs
1	$12 \cdot 1 = 12$
2	$12 \cdot 2 = 24$
3	$12 \cdot 3 = 36$
d	$12 \cdot d = 12d$

We multiply the number of dozen by 12 to find the number of eggs.

If d = the number of dozen eggs, the number of eggs is $12 \cdot d$, or, more simply, $12d$.

Self Check 8

Complete the table. Then use that information to write an algebraic expression that represents the number of yards in f feet.

Number of feet	Number of yards
3	
6	
9	
f	

Now Try **Problems 79 and 83**

ANSWERS TO SELF CHECKS

1. a. $t - 80$ **b.** $T + 6$ **c.** $(2a - 15)^2$ **2.** $m + 15$ **3.** p = the regular price of the sweater (in dollars); $p - 20$ = the sale price of the sweater (in dollars) **4.** x = the lottery payoff (in dollars); $\frac{x}{15}$ = each person's share (in dollars) **5.** x = the number of votes received by the challenger; $3x - 55$ = the number of votes received by the incumbent **6.** s = the amount donated to the scholarship fund (in dollars); $900 - s$ = the amount donated to the building fund (in dollars) **7.** x = the age of Bill Clinton; $x + 9$ = the age of Bill Cosby; $x - 9$ = the age of Bill Gates **8.** $1, 2, 3, \frac{f}{3}$

SECTION 1.7 STUDY SET

VOCABULARY

Fill in the blanks.

1. Variables and/or numbers can be combined with the operations of addition, subtraction, multiplication, and division to create algebraic __________.

2. A ________ is a letter (or symbol) that stands for a number.

3. A number, such as 8, is called a ________ because it does not change.

4. Phrases such as *increased by* and *more than* indicate the operation of ________. Phrases such as *decreased by* and *less than* indicate the operation of __________.

5. The word *product* indicates the operation of ____________. The word *quotient* indicates the operation of ________.

6. An _________ is a mathematical sentence that contains an = symbol. An algebraic _________ does not.

CONCEPTS

Classify each item as an algebraic expression or an equation.

7. a. $m + 18 = 23$ **b.** $m + 18$

8. a. $30x$ **b.** $30x = 600$

9. a. $\frac{c - 7}{5}$ **b.** $\frac{c - 7}{5} = 7c$

10. a. $r = \frac{2}{3}$ **b.** $\frac{2}{3}r$

11. What arithmetic operations does the expression $\frac{12 + 9t}{25}$ contain? What variable does it contain?

12. What arithmetic operations does the equation $4y - 14 = 5(6)$ contain? What variable does it contain?

13. a. Write an algebraic expression that is a combination of the number 10, the variable x, and the operation of addition.

b. Write an algebraic expression that is a combination of the numbers 3 and 2, the variable t, and the operations of multiplication and subtraction.

14. The illustration on the next page shows the commute to work (in miles) for two men, Mr. Lamb and Mr. Lopez, who work in the same office.

a. Who lives farther from the office?

b. How much farther?

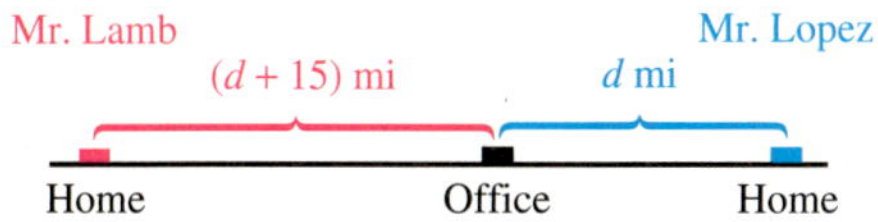

15. Match each algebraic expression to the correct phrase.

a. $c + 2$ i. twice c

b. $2 - c$ ii. c increased by 2

c. $c - 2$ iii. c less than 2

d. $2c$ iv. 2 less than c

16. Fill in the blank to complete the translation.

a. 16 less than m

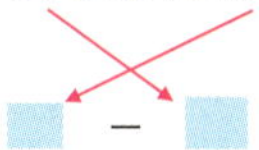

b. 16 is less than m

16 ☐ m

17. CUTLERY The knife shown below is 12 inches long. Write an algebraic expression that represents the length of the blade (in inches).

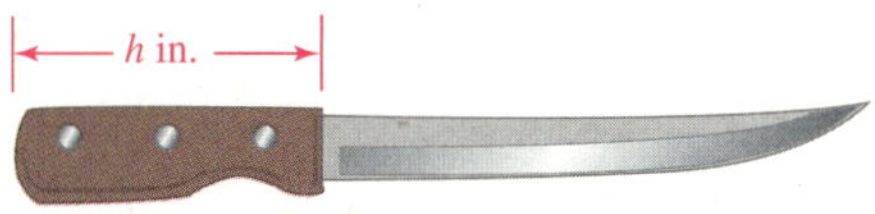

18. The following table shows the ages of three family members.

	Age (years)
Matthew	x
Sarah	$x - 8$
Joshua	$x + 2$

a. Who is the youngest person shown in the table?

b. Who is the oldest person listed in the table?

c. On whose age are the ages in the table based?

19. Complete the table. Then fill in the blank.

Number of decades	Number of years
1	
2	
3	
d	

We ________ the number of decades by 10 to find the number of years.

20. Complete the table. Then fill in the blank.

Number of inches	Number of feet
12	
24	
36	
i	

We ______ the number of inches by 12 to find the number of feet.

NOTATION

21. Write each algebraic expression in simpler form.

a. $x \cdot 8$ **b.** $5(t)$ **c.** $10 \div g$

d. $P \cdot r \cdot t$ **e.** $(x)(y)$

22. Consider the phrase:

the product of 5 and w increased by 30

Insert a comma in the phrase so that it translates to $5w + 30$.

GUIDED PRACTICE

Translate each phrase to an algebraic expression. If no variable is given, use x as the variable. See Example 1.

23. The sum of the length l and 15

24. The difference of a number and 10

25. The product of a number and 50

26. Three more than p

27. The amount won w less the amount lost l

28. The tax t added to c

29. P increased by twice p

30. 21 less than the total height h

31. The square of k, minus 2,005

32. s subtracted from S

33. 1 less than twice the attendance a

34. J reduced by 500

35. 1,000 split n equal ways

36. Exceeds the cost c by 25,000

37. 90 more than twice the current price p

38. 64 divided by the cube of y

39. 3 times the total of 35, h, and 300

40. Decrease x by -17

41. 680 fewer than the entire population p

42. Triple the number of expected participants

43. The product of d and 4, decreased by 15

44. The quotient of y and 6, cubed

45. Twice the sum of 200 and t

46. The square of the quantity 14 less than x

47. The absolute value of the difference of a and 2

48. The absolute value of a, decreased by 2

49. Ten times the distance d

50. Double the difference of x and 18

Write an algebraic expression that represents the unknown quantity. **See Example 2.**

51. GARDENING The height of a hedge was f feet before a gardener cut 2 feet off the top. Write an algebraic expression that represents the height of the trimmed hedge (in feet).

52. SHOPPING A married couple needed to purchase 21 presents for friends and relatives on their holiday gift list. If the husband purchased g presents, write an algebraic expression that represents the number of presents that the wife needs to buy.

53. PACKAGING A restaurant owner purchased s six-packs of cola. Write an algebraic expression that represents the number of cans that this would be.

54. NOISE The highest decibel reading during a rock concert was only 5 decibels shy of that of a jet engine. If a jet engine is normally j decibels, write an algebraic expression that represents the decibel reading for the concert.

55. SUPPLIES A pad of yellow legal paper contains p pages. If a lawyer uses 15 pages every day, write an algebraic expression that represents the number of days that one pad will last.

56. ACCOUNTING The projected cost c (in dollars) of a freeway was too low by a factor of 10. Write an algebraic expression that represents the actual cost of the freeway (in dollars).

57. RECYCLING A campus ecology club collected t tons of newspaper. A Boy Scout troop then contributed an additional 2 tons. Write an algebraic expression that represents the number of tons of newspaper that were collected by the two groups.

58. GRADUATION A graduating class of x people took buses that held 40 students each to an all-night graduation party. Write an algebraic expression that represents the number of buses that were needed to transport the class.

59. STUDYING A student will devote h hours to study for a government final exam. She wants to spread the studying evenly over a four-day period. Write an algebraic expression that represents the number of hours that she should study each day.

60. BASEBALL TEAMS After all c children complete a Little League tryout, the league officials decide that they have enough players for 8 teams of equal size. Write an algebraic expression that represents the number of players that will be on each team.

61. SCOTCH TAPE Suppose x inches of tape have been used off the roll shown in the next column. Write an algebraic expression that represents the number of inches of tape that are left on the roll.

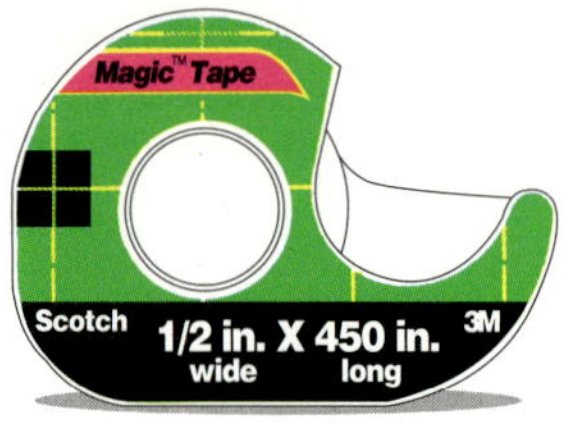

62. MODELING A model's skirt is x inches long. The designer then lets the hem down 2 inches. Write an algebraic expression that represents the length of the altered skirt (in inches).

In Problems 63–66, there are two unknowns. **See Examples 3 and 4.**

63. GEOMETRY The length of a rectangle is 6 inches longer than its width. Choose a variable to represent one of the unknown dimensions of the rectangle. Then write an algebraic expression that represents the other dimension.

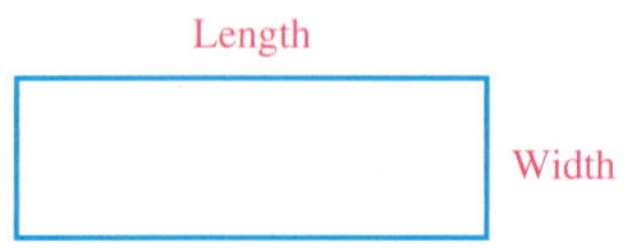

64. PLUMBING The smaller pipe shown below takes three times longer to fill the tank than does the larger pipe. Choose a variable to represent one of the unknown times it takes to fill the tank. Then write an algebraic expression that represents the other time.

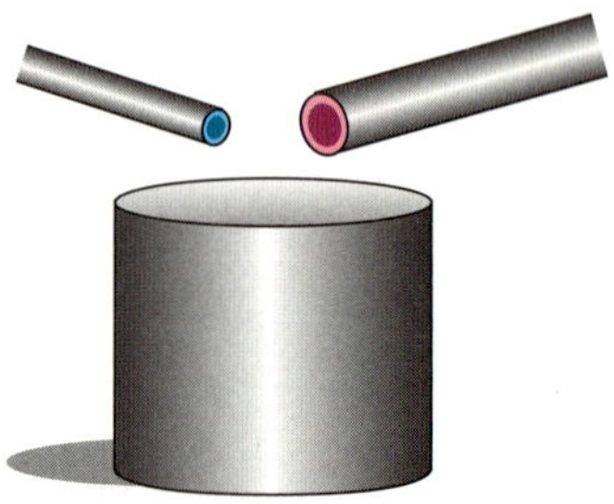

65. TRUCK REPAIR The truck radiator shown below was full of coolant. Then three quarts of coolant were drained from it. Choose a variable to represent one of the unknown amounts of coolant in the radiator. Then write an algebraic expression that represents the other amount.

66. SALE PRICES During a sale, the regular price of a CD was reduced by \$2. Choose a variable to represent one of the unknown prices of the CD. Then write an algebraic expression that represents the other price.

In Problems 59–62, there are two unknowns. See Example 5.

67. GEOGRAPHY Alaska is much larger than Vermont. To be exact, the area of Alaska is 380 square miles more than 50 times that of Vermont. Choose a variable to represent one area. Then write an algebraic expression that represents the other area.

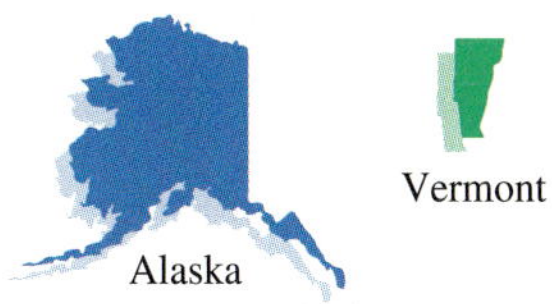

68. ROAD TRIPS On the second part of her trip, Tamiko drove 20 miles less than three times as far as the first part. Choose a variable to represent the number of miles driven on one part of her trip. Then write an algebraic expression that represents the number of miles driven on the other part.

69. DESSERTS The number of calories in a slice of pie is 100 more than twice the calories in a scoop of ice cream. Choose a variable to represent the number of calories in one type of dessert. Then write an algebraic expression that represents the number of calories in the other type.

70. WASTE DISPOSAL A waste disposal tank buried in the ground holds 15 gallons less than four times what a tank mounted on a truck holds. Choose a variable to represent the number of gallons that one type of tank holds. Then write an algebraic expression that represents the number of gallons the other tank holds.

In Problems 71–74, two approaches are used to represent the unknowns. See Example 6.

71. LANDSCAPING

a. Let b represent the height of the birch tree (in feet) that is shown in the next column. Write an algebraic expression that represents the height of the elm tree (in feet).

b. Let e stand for the height of the elm tree (in feet). Write an algebraic expression that represents the height of the birch tree (in feet).

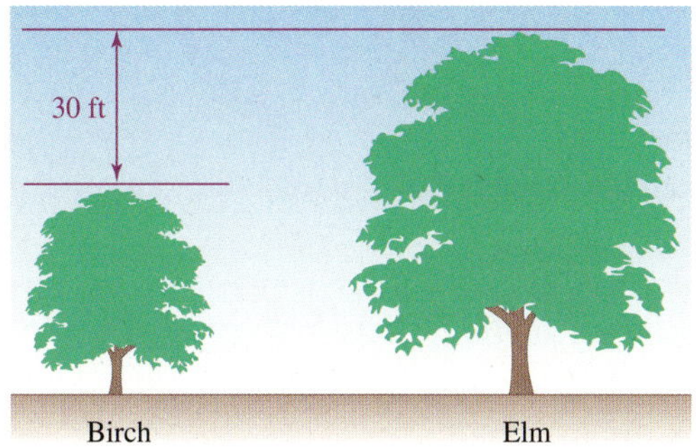

72. BUILDING MATERIALS

a. Let b = the length of the beam shown below (in feet). Write an algebraic expression that represents the length of the pipe.

b. Let p = the length of the pipe (in feet). Write an algebraic expression that represents the length of the beam.

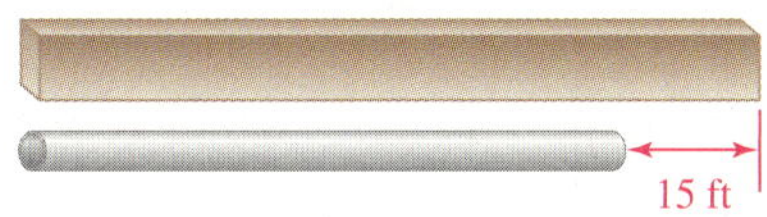

73. MARINE SCIENCE

a. Let s represent the length (in feet) of the great white shark shown below. Write an algebraic expression that represents the length (in feet) of the orca (killer whale).

b. Let w represent the length (in feet) of the orca (killer whale) shown below. Write an algebraic expression that represents the length (in feet) of the great white shark.

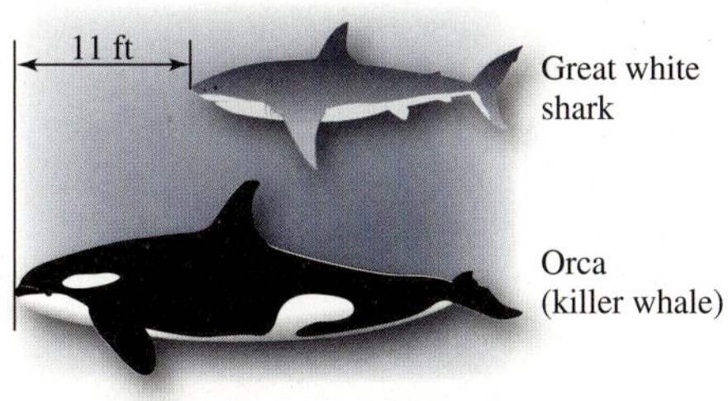

74. WEIGHTS AND MEASURES

a. Refer to the scale shown on the next page. Which mixture is heavier, A or B? How much heavier is it?

b. Let a represent the weight (in ounces) of mixture A. Write an algebraic expression that represents the weight (in ounces) of mixture B.

c. Let b represent the weight (in ounces) of mixture B. Write an algebraic expression that represents the weight (in ounces) of mixture A.

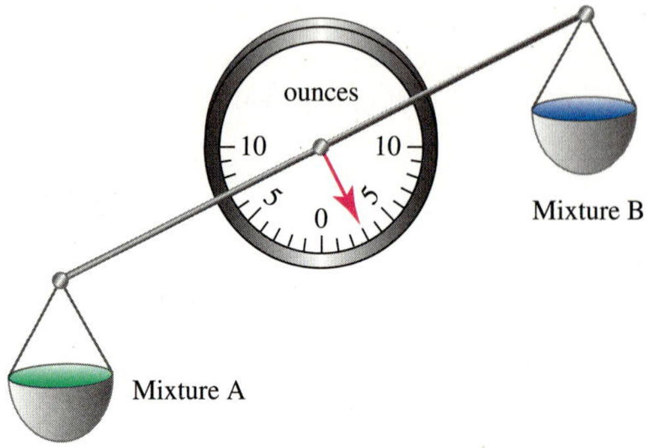

In Problems 75–78, there are three unknowns. **See Example 7.**

75. INVENTIONS The digital clock was invented 11 years before the automatic teller machine (ATM). The camcorder was invented 15 years after the ATM. Write algebraic expressions to represent the ages (in years) of each of these inventions. (Source: Wikipedia)

76. FAMOUS TOMS Tom Petty was born 6 years before Tom Hanks. Tom Cruise was born 6 years after Tom Hanks. Write algebraic expressions to represent the ages of each of these celebrities. (Source: celebritybirthdaylist.com)

77. NEW YORK ARCHITECTURE The Woolworth Building was completed 18 years before the Empire State Building. The United Nations Building was completed 21 years after the Empire State Building. Write algebraic expressions to represent the ages of each of these buildings. (Source: emporis.com)

78. CHILDREN'S BOOKS *The Tale of Peter Rabbit* was first published 24 years before *Winnie-the-Pooh. The Cat in the Hat* was first published 31 years after *Winnie-the-Pooh.* Write algebraic expressions to represent the ages of each of these books. (Source: Wikipedia)

Use a table to help answer Problems 79–86. **See Example 8.**

79. Write an algebraic expression that represents the number of seconds in m minutes.

80. Write an algebraic expression that represents the number of minutes in h hours.

81. Write an algebraic expression that represents the number of inches in f feet.

82. Write an algebraic expression that represents the number of feet in y yards.

83. Write an algebraic expression that represents the number of centuries in y years.

84. Write an algebraic expression that represents the number of decades in y years.

85. Write an algebraic expression that represents the number of dozen eggs in e eggs.

86. Write an algebraic expression that represents the number of days in h hours.

TRY IT YOURSELF

Translate each algebraic expression into words. (Answers may vary.)

87. $\frac{r}{3}$

88. $\frac{2}{d}$

89. $t - 50$

90. $c + 19$

91. xyz

92. $10ab$

93. $2m + 5$

94. $2s - 8$

95. A man sleeps x hours per day. Write an algebraic expression that represents

a. the number of hours that he sleeps in a week.

b. the number of hours that he sleeps in a year (non–leap year).

96. A store manager earns d dollars an hour. Write an algebraic expression that represents

a. the amount of money he will earn in an 8-hour day.

b. the amount of money he will earn in a 40-hour week.

97. A secretary earns an annual salary of s dollars. Write an algebraic expression that represents

a. her salary per month.

b. her salary per week.

98. Write an algebraic expression that represents the number of miles in f feet. (Hint: There are 5,280 feet in one mile.)

APPLICATIONS

99. ELECTIONS In 1960, John F. Kennedy was elected President of the United States with a popular vote only 118,550 votes more than that of Richard M. Nixon. Choose a variable to represent the number of votes received by one candidate. Then write an algebraic expression that represents the number of votes received by the other candidate.

100. THE BEATLES According to music historians, sales of the Beatles' second most popular single, *Hey Jude,* trail the sales of their most popular single, *I Want to Hold Your Hand,* by 2,000,000 copies. Choose a variable to represent the number of copies sold of one song. Then write an algebraic expression that represents the number of copies sold of the other song.

101. COMPUTER COMPANIES IBM was founded 80 years before Apple Computer. Dell Computer Corporation was founded 9 years after Apple. Let x represent the age (in years) of one of the companies. Write algebraic expressions to represent the ages (in years) of the other two companies.

102. VEHICLE WEIGHTS Refer to the illustration below. The car is 1,000 pounds lighter than the van. Choose a variable to represent the weight (in pounds) of one of the vehicles. Then write an algebraic expression that represents the weight (in pounds) of the other vehicle.

103. WITH/AGAINST THE WIND On a flight from Dallas to Miami, a jet airliner, which can fly 500 mph in still air, has a tail wind of x mph. The tail wind increases the speed of the jet. On the return flight to Dallas, the airliner flies into a head wind of the same strength. The head wind decreases the speed of the jet. Use this information to complete the table.

Wind conditions	Speed of jet (mph)
In still air	
With the tail wind	
Against the head wind	

104. SUB SANDWICHES Refer to the illustration in the next column. Write an algebraic expression that represents the length (in inches) of the second piece of the sandwich.

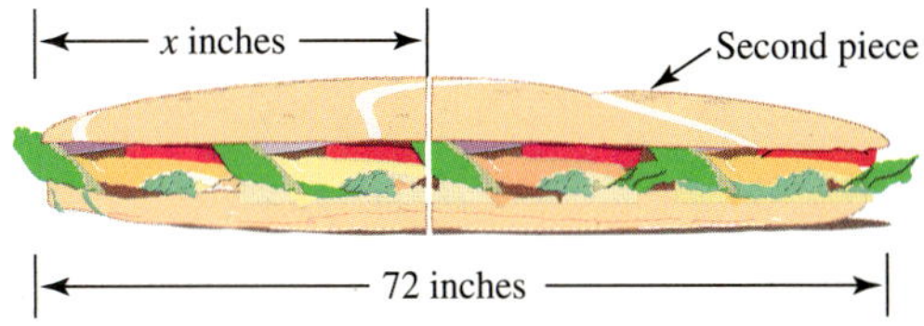

105. SAVINGS ACCOUNTS A student inherited \$5,000 and deposits x dollars in American Savings. Write an algebraic expression that represents the amount of money (in dollars) left to deposit in a City Mutual account.

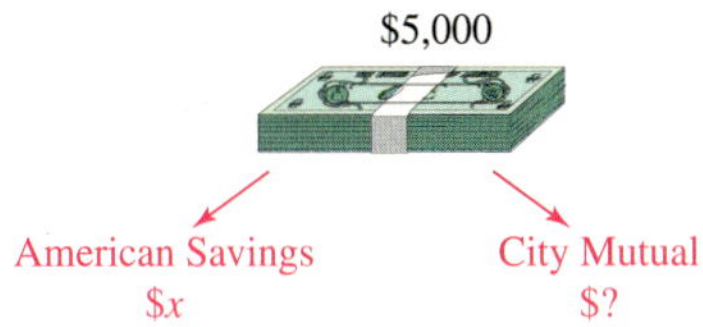

106. **a.** MIXING SOLUTIONS Solution 1 is poured into solution 2. Write an algebraic expression that represents the number of ounces in the mixture.

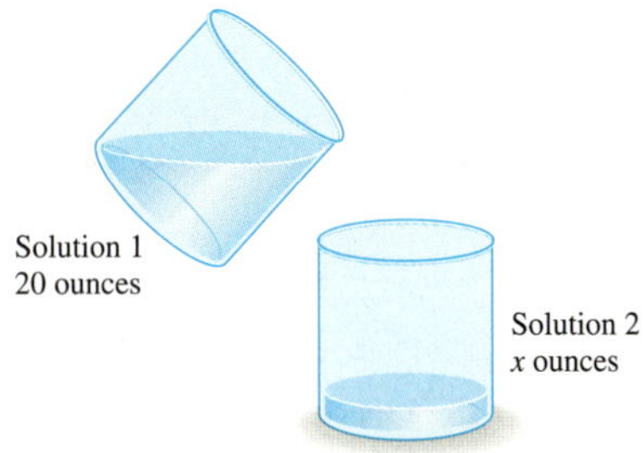

b. SNACKS Cashews were mixed with p pounds of peanuts to make 100 pounds of a mixture. Write an algebraic expression that represents the number of pounds of cashews that were used.

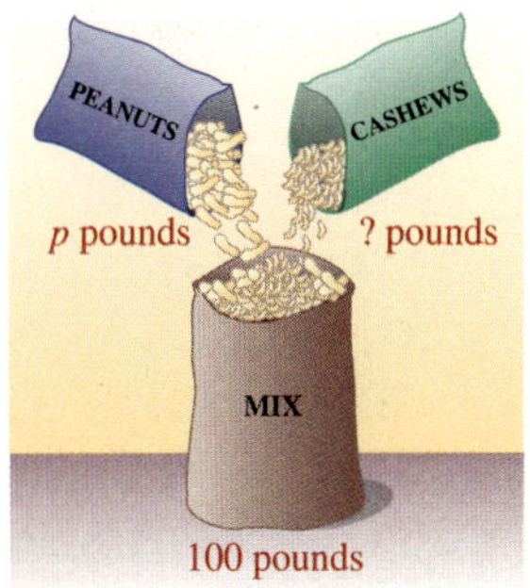

WRITING

107. Explain how variables are used in this section.

108. Explain the difference between the phrases *greater than* and *is greater than.*

109. Suppose in an application problem you were asked to find the unknown height of a building. Explain what is wrong with the following start.

Let $x =$ building

110. What is an algebraic expression?

SECTION 1.8

Applications Introduction: Variables and Formulas

A **variable** is a letter (or symbol) that stands for a number. The letters x, y, and z, are most commonly used as variables, but any letters can be used. For example, in the statement below about snowboarding, if we let c represent the number of calories burned and m represent the number of minutes snowboarding, the English words can be translated into a mathematical statement called an *equation.*

The number of calories burned	is	ten	times	the number of minutes snowboarding.
c	$=$	10	$\cdot$	m

In general, an **equation** is a mathematical sentence that contains an $=$ symbol. When verbs such as *is, are, gives,* and *yields* appear in problems, they often indicate an $=$ symbol. Equations such as $c = 10 \cdot m$, which express a relationship between two (or more) variables, are called **formulas.**

1. Translate each statement into an equation. (Answers may vary depending on the variables chosen.)

a. The sale price | is | $100 | minus | the discount.

b. The cost of dining out | equals | The cost of the meal | plus | $7 for parking.

c. 7 | times | the age of a dog in years | gives | the dog's equivalent human age.

d. The number of centuries | is | the number of years | divided by | 100.

e. The take-home pay | will be | $2,500 | minus | any deductions.

f. The number of decades | is | the number of years | divided by | 10.

g. The cost of renting the hall | is | $100 | times | the number of hours it is rented | plus | $200.

2. Some important mathematical formulas are given in words below. Write each formula using variables in place of the words. For example, the formula

Distance $=$ rate $\cdot$ time can be written $D = r \cdot t$

a. Profit = revenue − cost

b. Area = length · width

c. Diameter = 2 · radius

d. Regular price = sale price + discount

e. Perimeter = length + width + length + width

SECTION **1.8**

Evaluating Algebraic Expressions and Formulas

Objectives

1. Evaluate algebraic expressions.
2. Use formulas from business to solve application problems.
3. Use formulas from science to solve application problems.
4. Find the mean (average) of a set of values.

ARE YOU READY?

The following problems review some basic concepts that are important when working with algebraic expressions and formulas.

Evaluate each expression.

1. $5 \cdot 3^3 - 4 \cdot 2^3$

2. $8 - 3[5^2 - (7 - 3)^2]$

3. $\dfrac{13^2 - 5^2}{-3(5 - 3)^2}$

4. $\dfrac{2 \cdot 2^5 - 60 + (-4)}{5^4 - (-4)(-5)}$

Recall that an **algebraic expression** is a combination of variables and numbers with the operation symbols of addition, subtraction, multiplication, and division. In this section, we will be replacing the variables in algebraic expressions with numbers. Then, using the rule for the order of operations, we will *evaluate* each expression. We will also study formulas. Like algebraic expressions, formulas involve variables.

1 Evaluate algebraic expressions.

EXAMPLE 1 ***Plumbing*** The manufacturer's instructions for installing a kitchen garbage disposal are shown below.

a. Choose a variable to represent the length of one of the pieces of pipe (A, B, or C). Then write algebraic expressions to represent the lengths of the other two pieces.

b. Suppose model #201 is being installed. Find the length of each piece of pipe that is needed to connect the disposal to the drain line.

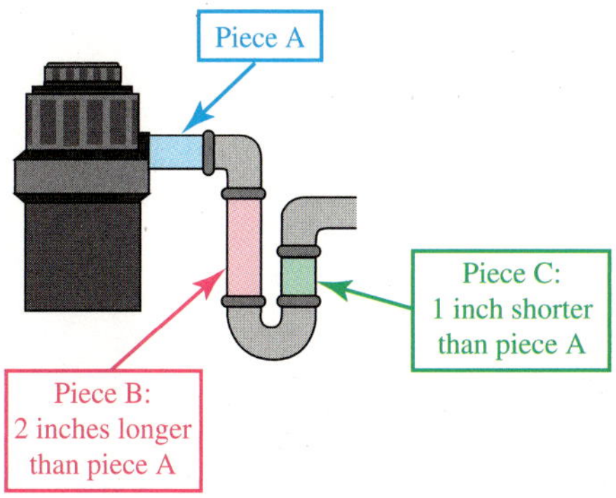

Model	Length of piece A
#101	2 inches
#201	3 inches
#301	4 inches

Self Check 1

PLUMBING Refer to Example 1. Suppose model #101 is being installed. Find the length of each piece of pipe that is needed to connect the disposal to the drain line.

Now Try **Problem 11**

Strategy There are three lengths of pipe to represent. We will begin by letting x = the length (in inches) of piece A.

WHY The lengths of the other pieces are related to (based on) the length of piece A.

Solution

a. Since the instructions call for piece B to be *2 inches longer* than piece A, and the length of piece A is represented by x, we have:

$x + 2$ = the length of piece B (in inches)

Since piece C is to be *1 inch shorter* than piece A,

$x - 1$ = the length of piece C (in inches)

The illustration on the right shows the algebraic expressions that can be used to represent the length of each piece of pipe.

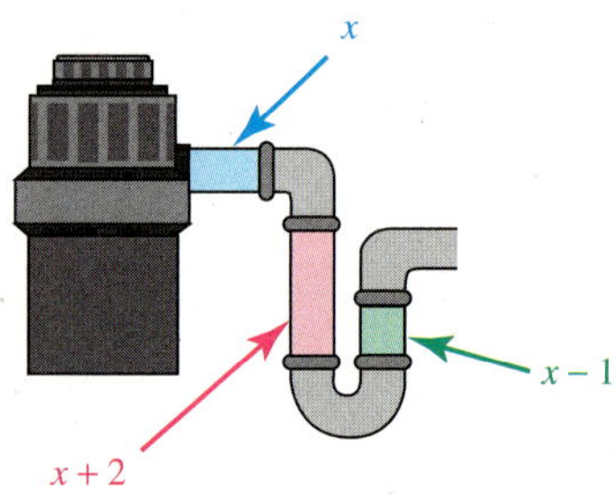

b. If model #201 is being installed, the table tells us that piece A should be 3 inches long. We can find the lengths of the other two pieces of pipe by replacing x with 3 in each of the algebraic expressions.

To find the length of piece B:

Replace x with 3.

$$x + 2 = 3 + 2$$
$$= 5$$

Piece B should be 5 inches long.

To find the length of piece C:

Replace x with 3.

$$x - 1 = 3 - 1$$
$$= 2$$

Piece C should be 2 inches long.

When we substitute given numbers for each of the variables in an algebraic expression and apply the order of operations rule, we are **evaluating the expression.** In the previous example, we say that we *substituted* 3 for x to evaluate the algebraic expressions $x + 2$ and $x - 1$.

Caution! When replacing a variable with its numerical value, we must often write the replacement number within parentheses to convey the proper meaning.

Self Check 2

Evaluate each expression for $y = 5$:

a. $5y - 4$

b. $\dfrac{-y - 15}{2}$

Now Try **Problems 15 and 17**

EXAMPLE 2 Evaluate each expression for $x = 3$:

a. $2x - 1$ **b.** $\dfrac{-x - 15}{6}$

Strategy We will replace x with the given value of the variable and evaluate the expression using the order of operations rule.

WHY To *evaluate an algebraic expression* means to find its numerical value, once we know the value of its variable.

Solution

a. $2x - 1 = 2(3) - 1$ Substitute 3 for x. Use parentheses.

$= 6 - 1$ Do the multiplication first: 2(3) = 6.

$= 5$ Do the subtraction.

b. $\dfrac{-x - 15}{6} = \dfrac{-(3) - 15}{6}$ Substitute 3 for x. Use parentheses. Don't forget to write the − sign in front of (3).

$= \dfrac{-3 - 15}{6}$ Simplify: −(3) = −3.

$= \dfrac{-3 + (-15)}{6}$ If it is helpful, write the subtraction of 15 as addition of the opposite of 15.

$$\begin{array}{r} 15 \\ +\ 3 \\ \hline 18 \end{array}$$

$= \dfrac{-18}{6}$ Do the addition: −3 + (−15) = −18.

$= -3$ Do the division.

EXAMPLE 3 Evaluate each expression for $a = -2$:

a. $4a^2 - 3a$ **b.** $-a + 3(1 + a)$ **c.** $a^3 - 5$

Strategy We will replace each a in the expression with the given value of the variable and evaluate the expression using the order of operations rule.

WHY To *evaluate an algebraic expression* means to find its numerical value, once we know the value of its variable.

Solution

a. $4a^2 - 3a = 4(-2)^2 - 3(-2)$ Substitute −2 for each a. Use parentheses.

$= 4(4) - 3(-2)$ Evaluate the exponential expression: $(-2)^2 = 4$.

$= 16 - (-6)$ Do each multiplication.

$= 16 + 6$ If it is helpful, write the subtraction of −6 as addition of the opposite of −6.

$= 22$ Do the addition.

$$\begin{array}{r} \overset{1}{1}6 \\ +\ 6 \\ \hline 22 \end{array}$$

b. $-a + 3(1 + a) = -(-2) + 3[1 + (-2)]$ Substitute −2 for each a. Use parentheses. Don't forget to write the − sign in front of (−2). Since another pair of grouping symbols are now needed, write brackets around 1 + (−2).

$= -(-2) + 3(-1)$ Do the addition within the brackets.

$= 2 + (-3)$ Simplify: −(−2) = 2. Do the multiplication: 3(−1) = −3.

$= -1$ Do the addition.

c. $a^3 - 5 = (-2)^3 - 5$ Substitute −2 for a. Use parentheses.

$= -8 - 5$ Evaluate the exponential expression: $(-2)^3 = -8$.

$= -8 + (-5)$ If it is helpful, write the subtraction of 5 as addition of the opposite of 5.

$= -13$ Do the addition.

Self Check 3

Evaluate each expression for $t = -3$:

a. $4t^2 - 2t$

b. $-t + 2(t + 1)$

c. $t^3 + 16$

Now Try **Problems 19, 21, and 23**

To evaluate algebraic expressions containing two or more variables, we need to know the value of each variable.

Self Check 4

Evaluate each expression for $r = -1$ and $s = 5$:

a. $(5rs + 4s)^2$

b. $|-8s - 2r|$

Now Try **Problems 27 and 29**

EXAMPLE 4 Evaluate each expression for $h = -1$ and $g = 5$:

a. $(8hg + 6g)^2$ **b.** $|-5g - 7h|$

Strategy We will replace each h and g in the expression with the given value of the variable and evaluate the expression using the order of operations rule.

WHY To *evaluate an expression* means to find its numerical value, once we know the values of its variables.

Solution

a. $(8hg + 6g)^2 = [8(-1)(5) + 6(5)]^2$ Substitute −1 for h and 5 for g. Use parentheses. Since another pair of grouping symbols are now needed, write brackets around 8(−1)(5) + 6(5).

$= (-40 + 30)^2$ Do the multiplication within the brackets.

$= (-10)^2$ Do the addition within the parentheses.

$= 100$ Evaluate the exponential expression: $(-10)^2 = 100$.

b. $|-5g - 7h| = |-5(5) - 7(-1)|$ Substitute −1 for h and 5 for g. Use parentheses.

$= |-25 - (-7)|$ Do the multiplication within the absolute value symbols.

$= |-25 + 7|$ If it is helpful, write the subtraction of −7 as the addition of the opposite of −7.

$= |-18|$ Do the addition.

$= 18$ Find the absolute value of −18.

$$\begin{array}{r} \overset{1\,15}{\cancel{2}\cancel{5}} \\ -\ 7 \\ \hline 18 \end{array}$$

2 Use formulas from business to solve application problems.

A **formula** is an equation that states a mathematical relationship between two or more variables. Formulas are used in many fields: economics, physical education, biology, automotive repair, and nursing, just to name a few. In this section, we will consider several formulas from business, science, and mathematics.

A formula to find the sale price

If a car that usually sells for \$22,850 is discounted \$1,500, you can find the **sale price** using the formula

Sale price = original price − discount

Using the variables s to represent the sale price, p the original price, and d the discount, we can write this formula as

$$\boxed{s = p - d}$$

To find the sale price of the car, we substitute 22,850 for p, 1,500 for d, and evaluate the right side of the equation.

$s = p - d$ — This is the sale price formula.

$= 22{,}850 - 1{,}500$ — Substitute 22,850 for p and 1,500 for d.

$= 21{,}350$ — Do the subtraction.

$$\begin{array}{r} 22{,}850 \\ -\ 1{,}500 \\ \hline 21{,}350 \end{array}$$

The sale price of the car is $21,350.

A formula to find the retail price

To make a profit, a merchant must sell a product for more than he paid for it. The price at which he sells the product, called the **retail price,** is the *sum* of what the item cost him and the markup.

Retail price = cost + markup

Using the variables r to represent the retail price, c the cost, and m the markup, we can write this formula as

$$\boxed{r = c + m}$$

As an example, suppose that a store owner buys a lamp for $35 and then marks up the cost $20 before selling it. We can find the retail price of the lamp using this formula.

$r = c + m$ — This is the retail price formula.

$= 35 + 20$ — Substitute 35 for c and 20 for m.

$= 55$ — Do the addition.

The retail price of the lamp is $55.

A formula to find profit

The **profit** a business makes is the *difference* of the revenue (the money it takes in) and the costs.

Profit = revenue − costs

Using the variables p to represent the profit, r the revenue, and c the costs, we have the formula

$$\boxed{p = r - c}$$

EXAMPLE 5 ***Films*** It cost Universal Studios about $523 million to make and distribute the film *Jurassic Park*. If the studio has received approximately $920 million to date in worldwide revenue from the film, find the profit the studio has made on this movie. (Source: swivel.com)

Strategy To find the profit, we will substitute the given values in the formula $p = r - c$ and evaluate the expression on the right side of the equation.

WHY The variable p represents the unknown profit.

Self Check 5

FILMS It cost Paramount Pictures about $394 million to make and distribute the film *Forrest Gump*. If the studio has received approximately $679 million to date in worldwide revenue from the film, find the profit the studio has made on this movie. (Source: swivel.com)

Now Try **Problem 35**

Solution The studio has received \$920 million in revenue r and the cost c to make and distribute the movie was \$523 million. To find the profit p, we proceed as follows.

$p = r - c$ This is the formula for profit.

$= 920 - 523$ Substitute 920 for r and 523 for c. The units are millions of dollars.

$= 397$ Do the subtraction.

$$\begin{array}{r} \overset{8\,\overset{11}{1}\,10}{920} \\ -523 \\ \hline 397 \end{array}$$

Universal Studios has made \$397 million in profit on the film *Jurassic Park*.

3 Use formulas from science to solve application problems.

A formula to find the distance traveled

If we know the rate (speed) at which we are traveling and the time we will be moving at that rate, we can find the distance traveled using the formula

Distance = rate · time

Using the variables d to represent the distance, r the rate, and t the time, we have the formula

$$\boxed{d = rt}$$

Self Check 6

SPEED LIMITS Nevada's speed limit for trucks on rural interstate highways is 75 mph. How far would a truck travel in 3 hours at that speed?

Now Try **Problem 39**

EXAMPLE 6 ***Interstate Speed Limits*** Three state speed limits for trucks are shown below. At each of these speeds, how far would a truck travel in 3 hours?

Strategy To find the distance traveled, we will substitute the given values in the formula $d = rt$ and evaluate the expression on the right side of the equation.

WHY The variable d represents the unknown distance traveled.

Solution To find the distance traveled by a truck in Oregon, we write

$d = rt$ This is the formula for distance traveled.

$= 55(3)$ Substitute: 55 mph is the rate r and 3 hours is the time t.

$= 165$ Do the multiplication. The units of the answer are miles.

$$\begin{array}{r} \overset{1}{5}5 \\ \times\ 3 \\ \hline 165 \end{array}$$

At 55 mph, a truck would travel 165 miles in 3 hours. We can use a table to display the calculations for each state.

	r ·	t =	d
Oregon	55	3	165
Michigan	60	3	180
Virginia	65	3	195

This column gives the distance traveled, in miles.

$$\begin{array}{r} 60 \\ \times\ 3 \\ \hline 180 \end{array} \qquad \begin{array}{r} \overset{1}{6}5 \\ \times\ 3 \\ \hline 195 \end{array}$$

Caution! When using $d = rt$ to find distance, make sure that the units are similar. For example, if the rate is given in miles per hour, the time must be expressed in hours.

A formula to find the distance an object falls

The distance an object falls (in feet) when it is dropped from a height is related to the time (in seconds) that it has been falling by the formula

$$\text{Distance fallen} = 16 \cdot (\text{time})^2$$

Using the variables d to represent the distance and t the time, we have

$$\boxed{d = 16t^2}$$

EXAMPLE 7 ***Balloon Rides*** Find the distance a camera fell in 6 seconds if it was dropped overboard by a vacationer taking a hot-air balloon ride.

Strategy To find the distance the camera fell, we will substitute the given time in the formula $d = 16t^2$ and evaluate the expression on the right side of the equation.

WHY The variable d represents the distance fallen.

Solution

$d = 16t^2$	This is the formula for distance fallen.
$= 16(6)^2$	The camera fell for 6 seconds. Substitute 6 for t.
$= 16(36)$	Evaluate the exponential expression: $6^2 = 36$.
$= 576$	Do the multiplication.

$$\begin{array}{r} 36 \\ \times\ 16 \\ \hline 216 \\ 360 \\ \hline 576 \end{array}$$

The camera fell 576 feet.

Self Check 7

FREEFALL Find the distance a rock fell in 3 seconds if it was dropped over the edge of the Grand Canyon.

Now Try **Problem 43**

4 Find the mean (average) of a set of values.

The **mean,** or **average,** of a set of numbers is a value around which the numbers are grouped. To find the mean, we divide the *sum* of all the values by the *number* of values. Writing this as a formula, we get

$$\text{Mean} = \frac{\text{sum of the values}}{\text{number of values}}$$

Using the variables S to represent the sum and n the number of values, we have

$$\boxed{\text{Mean} = \frac{S}{n}}$$

EXAMPLE 8 ***Response Time*** A police department recorded the length of time between incoming 911 calls and the arrival of a police unit at the scene. The response times (in minutes) for one 24-hour period are listed below. Find the mean (average) response time.

Response times

5 min	3 min	6 min	2 min	7 min	4 min	3 min	2 min

Self Check 8

WEB TRAFFIC The number of hits a website received each day for one week are listed below.
Mon: 392, Tues: 931,
Wed: 842, Thurs: 566,
Fri: 301, Sat: 103, Sun: 43
Find the mean (average) number of hits each day.

Now Try **Problem 47**

Strategy We will count the number of response times and calculate their sum.

WHY To find the mean of a set of values, we divide the sum of the values by the number of values.

Solution There are 8 response times. To find their sum, it is helpful to look for groups of numbers that add to 10.

$$5 + 3 + 6 + 2 + 7 + 4 + 3 + 2 = 32$$

$5 + 3 + 2 = 10$
$6 + 4 = 10$
$7 + 3 = 10$

Now we use the formula to find the mean.

$$\text{Mean} = \frac{S}{n}$$ This is the formula to find the mean (average).

$$= \frac{32}{8}$$ Substitute 32 for S, the sum of the response times. Substitute 8 for n, the number of response times.

$$= 4$$ Do the division.

The mean response time was 4 minutes.

When a value in a set appears more than once, that value has a greater "influence" on the mean than another value that only occurs a single time. To simplify the process of finding a mean, any value that appears more than once can be "weighted" by multiplying it by the number of times it occurs. A mean that is found in this way is called a **weighted mean**.

Self Check 9

On the first question of an instructor's evaluation, 14 students marked 1 for strongly agree, 10 students marked 2 for agree, 6 students marked 3 for disagree, and 4 students marked 4 for strongly disagree. What was the average (mean) response for the first question on the evaluation?

Now Try **Problems 94, 96**

EXAMPLE 9 ***Hotel Reservations.*** In an effort to improve customer service, a hotel electronically recorded the number of times the reservation desk telephone rang before it was answered by a receptionist. The results of the week-long survey are shown in the table. Find the mean (average) number of times the phone rang before a receptionist answered.

Number of rings	Number of calls
1	11
2	46
3	45
4	28
5	20

Strategy First, we will determine the total number of times the reservation desk telephone rang during the week. Then we will divide that result by the total number of calls received.

WHY To find the mean of a set of values, we divide the sum of the values by the number of values.

Solution

To find the total number of rings, we multiply each *number of rings* (1, 2, 3, 4, and 5 rings) by the respective number of occurrences and add those subtotals.

$$\text{Total number of rings} = 11(1) + 46(2) + 45(3) + 28(4) + 20(5)$$

The total number of calls received was $11 + 46 + 45 + 28 + 20$. To find the mean, we divide the total number of rings by the total number of calls.

$$\text{Mean} = \frac{11(1) + 46(2) + 45(3) + 28(4) + 20(5)}{11 + 46 + 45 + 28 + 20}$$

$$= \frac{11 + 92 + 135 + 112 + 100}{150}$$ In the numerator, do the multiplications. In the denominator, do the additions.

$$= \frac{450}{150}$$ Do the addition.

$$= 3$$ Simplify the fraction.

The mean number of times the phone rang before it was answered was 3.

THINK IT THROUGH *Study Time*

"Your success in school is dependent on your ability to study effectively and efficiently. The results of poor study skills are wasted time, frustration, and low or failing grades."

Effective Study Skills, Dr. Bob Kizlik, 2004

For a course that meets for h hours each week, the formula $H = 2h$ gives the suggested number of hours H that a student should study the course outside of class each week. If a student expects difficulty in a course, the formula can be adjusted upward to $H = 3h$. Use the formulas to complete the table on the right.

	Suggested study time (hours per week)	Expanded study time (hours per week)
If a course meets for:		
2 hours per week		
3 hours per week		
4 hours per week		
5 hours per week		

ANSWERS TO SELF CHECKS

1. piece B: 4 in.; piece C: 1 in. **2. a.** 21 **b.** −10 **3. a.** 42 **b.** −1 **c.** −11 **4. a.** 25 **b.** 38 **5.** Paramount Pictures has made $285 million in profit on the movie *Forrest Gump.* **6.** 225 mi **7.** 144 ft **8.** 454 hits **9.** 2

SECTION 1.8 STUDY SET

VOCABULARY

Fill in the blanks.

1. An algebraic ________ is a combination of variables, numbers, and the operation symbols for addition, subtraction, multiplication, and division.
2. When we substitute 5 for x in the algebraic expression $7x + 10$ and apply the order of operations rule, we are ________ the expression.
3. To evaluate $a^2 + 10a + 1$ for $a = -3$, we ________ -3 for a and apply the order of operations rule.
4. A _______ is an equation that states a mathematical relationship between two or more variables.
5. The rule for the _____ of operations guarantees that an evaluation of a numerical expression will result in a single answer.
6. To find the _____ (or average) of a set of values, we divide the sum of the values by the number of values.

CONCEPTS

7. Use variables to write the formula that relates each of the quantities listed below.
   a. Sale price, original price, discount
   b. Profit, revenue, costs
   c. Retail price, cost, markup
8. Use variables to write the formula that relates each of the quantities listed below.
   a. Distance, rate, time
   b. The distance an object falls when dropped, time
   c. Mean, number of values, sum of values

NOTATION

9. Complete the solution. Evaluate the expression for $a = 5$.

$$\begin{aligned} 9a - a^2 &= 9(\square) - (5)^2 \\ &= 9(5) - \square \\ &= \square - 25 \\ &= 20 \end{aligned}$$

10. ***Complete each step. Evaluate the expression for $x = -2$, and $y = 4$.***

$$\begin{aligned} -x + 6y &= -(\square) + 6(\square) \\ &= \square + 24 \\ &= 26 \end{aligned}$$

GUIDED PRACTICE

In Problems 11–14, write algebraic expressions to represent the three unknowns and then evaluate each of them for the given value of the variable. See Example 1.

11. PLAYGROUND EQUIPMENT The plans for building a children's swing set are shown below.

a. Choose a variable to represent the length (in inches) of one part of the swing set. Then write algebraic expressions that represent the lengths (in inches) of the other two parts.

b. If the builder chooses to have part 1 be 60 inches long, how long should parts 2 and 3 be?

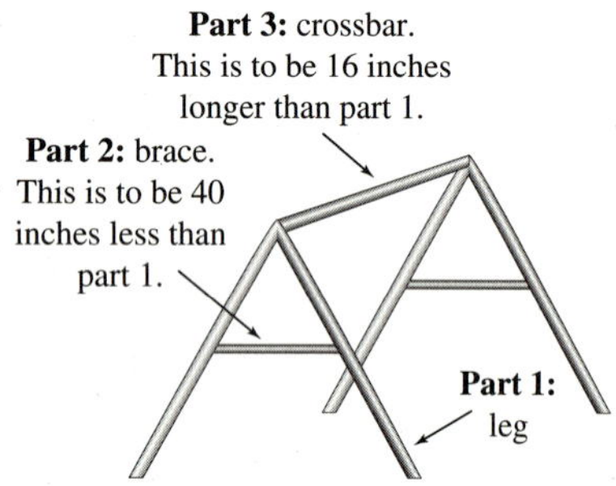

12. ART DESIGN A television studio art department plans to construct two sets of decorations out of plywood, using the plan shown below.

a. Choose a variable to represent the height (in inches) of one piece of plywood. Then write algebraic expressions that represent the heights (in inches) of the other two pieces.

b. Designers will make the first set of three pieces for the foreground. Piece A will be 15 inches high. How high should pieces B and C be?

c. Designers will make another set of three pieces for the background. Piece A will be 30 inches high. How high should pieces B and C be?

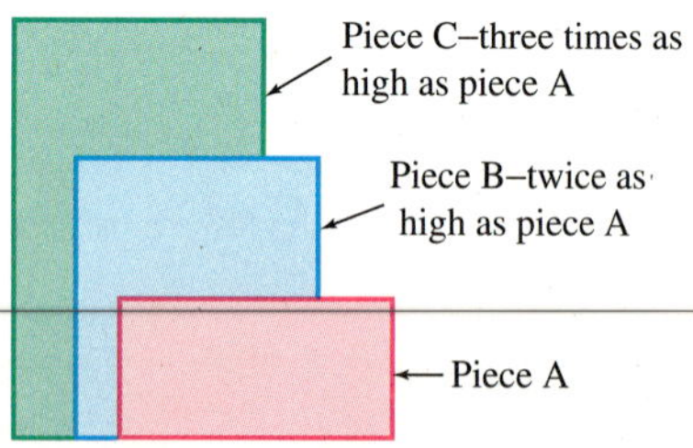

13. VEHICLE WEIGHTS An H2 Hummer weighs 340 pounds less than twice a Honda Element. A Smart Fortwo car weighs 1,720 pounds less than a Honda Element.

a. Choose a variable to represent the weight (in pounds) of one car. Then write algebraic expressions that represent the weights (in pounds) of the other two cars.

b. If the weight of the Honda Element is 3,370 pounds, find the weights of the other two cars.

14. BATTERIES An AAA-size battery weighs 53 grams less than a C-size battery. A D-size battery weighs 5 grams more than twice a C-size battery.

a. Choose a variable to represent the weight (in grams) of one size battery. Then write algebraic expressions that represent the weights (in grams) of the other two batteries.

b. If the weight of a C-size battery is 65 grams, find the weights of the other two batteries.

Evaluate each expression for the given value of the variable. See Example 2.

15. $10x - 3$ for $x = 3$

16. $4a - 2$ for $a = 9$

17. $\dfrac{-n - 1}{3}$ for $n = 11$

18. $\dfrac{-b - 2}{7}$ for $b = 5$

Evaluate each expression for the given value of the variable. See Example 3.

19. $3x^2 - 2x$ for $x = -2$

20. $4n^2 - 5n$ for $n = -3$

21. $-y + 3(1 + y)$ for $y = -10$

22. $-b + 6(2 + b)$ for $b = -8$

23. $h^3 - 24$ for $h = -3$

24. $t^3 - 30$ for $t = -4$

25. $n^4 + n^2$ for $n = -1$

26. $d^4 + d^3$ for $d = -2$

Evaluate each expression for the given values of the variables. See Example 4.

27. $(2ab + 4b)^2$ for $a = -5$ and $b = 2$

28. $(3xy + 2y)^2$ for $x = -4$ and $y = 3$

29. $|-6r - 8s|$ for $r = -11$ and $s = 9$

30. $|-7t - 10x|$ for $t = -12$ and $x = 15$

Use the correct formula to solve each problem. See Objective 2 and Example 5.

31. SPORTING GOODS Find the sale price of a pair of skis that usually sells for \$200 but is discounted \$35.

32. OFFICE FURNISHINGS If a desk chair that usually sells for \$199 is discounted \$38, what is the sale price of the chair?

33. CLOTHING STORES A store owner buys a pair of pants for \$125 and marks them up \$65 for sale. What is the retail price of the pants?

34. SNACKS It costs a snack bar owner 20 cents to make a snow cone. If the markup is 50 cents, what is the retail price of a snow cone?

35. SMALL BUSINESSES On its first night of business, a pizza parlor brought in \$445. The owner estimated his costs that night to be \$295. What was the profit?

36. FLORISTS For the month of June, a florist's cost of doing business was \$3,795. If June revenues totaled \$5,115, what was her profit for the month?

37. FUNDRAISERS A school carnival brought in revenues of \$13,500 and had costs of \$5,300. What was the profit?

38. PRICING A shopkeeper marks up the cost of every item she carries by the amount she paid for the item. If a fan costs her \$27, what does she charge for the fan?

Use the correct formula to solve each problem. See Example 6.

39. AIRLINES Find the distance covered by a jet if it travels for 3 hours at 550 mph.

40. ROAD TRIPS Find the distance covered by a car traveling 60 miles per hour for 5 hours.

41. HIKING A hiker can cover 12 miles per day. At that rate, how far will the hiker travel in 8 days?

42. TURTLES A turtle can walk 250 feet per minute. At that rate, how far can a turtle walk in 5 minutes?

Use the correct formula to solve each problem. See Example 7.

43. FREE FALL Find the distance a ball has fallen 2 seconds after being dropped from a tall building.

44. SIGHTSEEING A visitor to the Grand Canyon accidently dropped her sunglasses over the edge. It took 9 seconds for the sunglasses to fall directly to the bottom of the canyon. How far above the canyon bottom was she standing?

45. BRIDGE REPAIR A steel worker dropped his wrench while tightening a cable on the top of a bridge. It took 4 seconds for the wrench to fall straight to the ground. How far above ground level was the man working?

46. LIGHTHOUSES An object was dropped from the top of the Tybee Island Lighthouse (located near Savanna, Georgia). It took 3 seconds for the object to hit the ground. How tall is the lighthouse?

Use the correct formula to find each mean (average). See Example 8.

47. BOWLING Find the mean score for a bowler who rolled scores of 254, 225, and 238.

48. YAHTZEE A player had scores of 288, 192, 264, and 124 at a Yahtzee tournament. What was his mean score?

49. FISHING The weights of each of the fish caught by those on a deep-sea fishing trip are listed below. What was the mean weight?

23 lb 18 lb 37 lb 11 lb 18 lb 26 lb 42 lb 25 lb

50. GRADES Find the mean score of the following test scores: 76, 83, 79, 91, 0, 73.

TRY IT FOR YOURSELF

Evaluate each expression for the given value(s) of the variable(s).

51. $\frac{x-8}{2}$ for $x = -4$

52. $\frac{-10+y}{-4}$ for $y = -6$

53. $-p$ for $p = -4$

54. $-j$ for $j = -9$

55. $2(p+9) + 2p$ for $p = -12$

56. $3(r-20) + 2r$ for $r = 15$

57. $x^2 - x - 7$ for $x = -5$

58. $a^2 + 3a - 9$ for $a = -3$

59. $\frac{x-y}{a-b}$ for $x = -1, y = 8, a = 6$, and $b = 3$

60. $\frac{m-n}{c-d}$ for $m = -20, n = -40, c = -5$, and $d = -10$

61. $\frac{-b^2+3b}{2b+1}$ for $b = -4$

62. $\frac{-a^2+5a}{2a+12}$ for $a = -3$

63. $\frac{24+k}{3k}$ for $k = 3$

64. $\frac{4-h}{h-4}$ for $h = -1$

65. $(x-a)^2 + (y-b)^2$ for $x = -2, y = 1, a = 5$, and $b = -3$

66. $2a^2 + 2ab + b^2$ for $a = -5$ and $b = -1$

67. $|6-x|$ for $x = 50$

68. $|3c-1|$ for $c = -1$

69. $-2|x|-7$ for $x = -7$

70. $|x^2 - 7^2|$ for $x = 7$

71. $2 - [10 - x(5h-1)]$ for $x = -2$ and $h = 2$

72. $1 - [8 - c(2k-7)]$ for $c = -3$ and $k = 4$

73. $b^2 - 4ac$ for $b = -3, a = 4$, and $c = -1$

74. $3r^2h$ for $r = 4$ and $h = 2$

75. $\frac{x}{y+10}$ for $x = 30$ and $y = -10$

76. $\frac{e}{3f+24}$ for $e = 24$ and $f = -8$

77. $\frac{50-6s}{-t}$ for $s = 5$ and $t = 4$

78. $\dfrac{7v - 5r}{-r}$ for $v = 8$ and $r = 4$

79. $5rs^2t$ for $r = 2$, $s = -3$, and $t = -3$

80. $-3bk^2t$ for $b = -5$, $k = -2$, and $t = -3$

81. $\dfrac{|a^2 - b^2|}{2a - b}$ for $a = -2$ and $b = -5$

82. $\dfrac{-|2x - 3y + 10|}{-3 - y}$ for $x = 0$ and $y = -4$

APPLICATIONS

83. ACCOUNTING Refer to the financial statement for Avon Products, Inc., shown below. Find the operating profit for the year ending January 2008 and the year ending January 2009.

Annual Financials: Income Statement
(All dollar amounts in millions)

	Year ending Jan. '08	Year ending Jan. '09
Total revenues	9,939	10,690
Cost of goods sold	3,773	3,946
Operating profit		

(Source: *Business Week*)

84. DISTANCE TRAVELED Complete the table below by finding the distance traveled in each instance.

	Rate (mph)	· time (hr)	= distance (mi)
Bike	12	4	
Walking	3	2	
Car	3	x	

85. DASHBOARDS The illustration below shows part of a dashboard. Explain what each of the three instruments measures. What is the formula that mathematically relates these measurements?

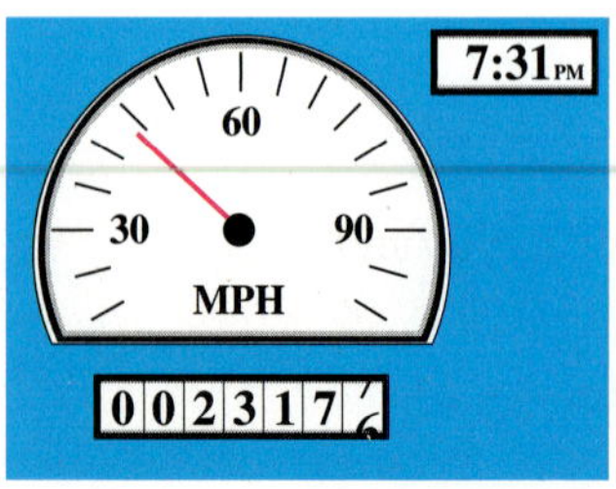

86. SPREADSHEETS A store manager wants to use a spreadsheet to post the prices of items on sale. If column B in the following table lists the regular price and column C lists the discount, write a formula using column names to have a computer find the sale price to print in column D. Then fill in column D with the correct sale price.

	A	B	C	D
1	Bath towel set	\$25	\$5	
2	Pillows	\$15	\$3	
3	Comforter	\$53	\$11	

87. DEALER MARKUPS A car dealer marks up the cars he sells \$500 above factory invoice (that is, \$500 over what it costs him to purchase the car from the factory).

a. Complete the following table.

Model	Factory invoice (\$)	Markup (\$)	Price (\$)
Minivan	25,600		
Pickup	23,200		
Convertible	x		

b. Write a formula that represents the price p of a car if the factory invoice is f dollars.

88. FALLING OBJECTS See the table below. First, find the distance in feet traveled by a falling object in 1, 2, 3, and 4 seconds. Enter the results in the middle column. Then find the distance the object traveled over each time interval and enter it in the right column.

Time falling	Distance traveled (ft)	Time intervals
1 sec		Distance traveled from 0 sec to 1 sec
2 sec		Distance traveled from 1 sec to 2 sec
3 sec		Distance traveled from 2 sec to 3 sec
4 sec		Distance traveled from 3 sec to 4 sec

89. DISTANCE TRAVELED

a. When in orbit, the space shuttle travels at a rate of approximately 17,250 miles per hour. How far does it travel in one day?

b. The speed of light is approximately 186,000 miles per second. How far will light travel in 1 minute?

c. The speed of a sound wave in air is about 1,100 feet per second at normal temperatures. How far does it travel in half a minute?

90. SOD FARMS. The expression $20{,}000 - 3s$ gives the number of square feet of sod that are left in a field after s strips have been removed. Suppose a city orders 7,000 strips of sod. Evaluate the expression and explain the result.

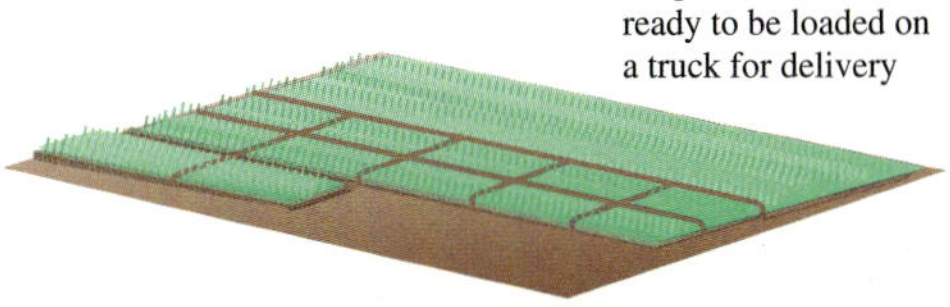

Strips of sod, cut and ready to be loaded on a truck for delivery

91. CHEMISTRY A compound being studied by chemists is stored at a temperature of −5°C. At 9:20 a.m., they began an experiment that produced a chemical reaction in the compound that increased its temperature 2°C per minute. The chemical reaction ended at 9:52 a.m.

a. Complete the table below that gives the temperature of the compound at several times during the experiment.

Time	Temperature °C
9:20 a.m.	
9:21 a.m.	
9: 22 a.m.	
9:30 a.m.	
9:45 a.m.	

b. Let C = the temperature of the compound in °C and m = the number of minutes after 9:20 a.m. Complete the formula below that finds the compound's temperature at any time between 9:20 a.m. and 9:52 a.m.

$C = \square + \square$

c. How many minutes after 9:20 a.m. is 9:52 a.m.?

d. Use your answers to parts b and c to find the temperature of the compound at 9:52 a.m.

92. AUTO INSURANCE See the premium comparison in the table. What is the average 6-month insurance premium?

Allstate	$2,672	Mercury	$1,370
Auto Club	$1,680	State Farm	$2,737
Farmers	$2,485	20th Century	$1,692

Criteria: Six-month premium. Husband, 45, drives a 1995 Explorer, 12,000 annual miles. Wife, 43, drives a 1996 Dodge Caravan, 12,000 annual miles. Son, 17, is an occasional operator. All have clean driving records.

93. TRANSPORTATION SECURITY OFFICER To determine the average afternoon wait time in security lines at an airport, officials monitored four passengers, each at a different gate. The time that each passenger entered a security line and the time the same passenger cleared the checkpoint was recorded, as shown below. Find the average (mean) wait time for these passengers.

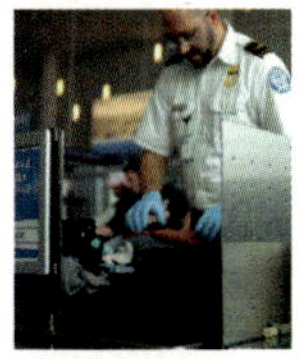

	Time entered	Time cleared
Passenger at Gate A	3:05 pm	3:21 pm
Passenger at Gate B	3:03 pm	3:13 pm
Passenger at Gate C	3:01 pm	3:09 pm
Passenger at Gate D	3:02 pm	3:16 pm

94. YOUTUBE VIDEO CONTEST A video contest is to be part of a promotional kickoff for a new sports drink. The prizes to be awarded are shown.

YouTube Video Contest

Grand prize: Disney World vacation plus $2,500

Four 1st place prizes of $500
Thirty-five 2nd place prizes of $150
Eighty-five 3rd place prizes of $25

a. How much money will be awarded in the promotion?

b. What is the mean (average) cash prize?

95. ENERGY USAGE Refer to the illustration on the next page. Find the mean number of therms of natural gas used per month.

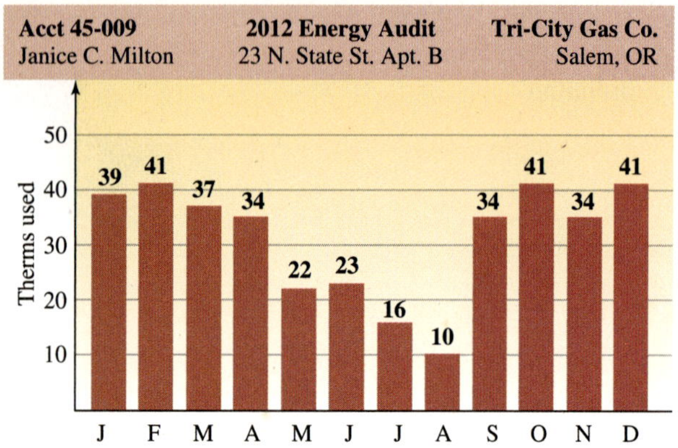

96. CUSTOMER SATISFACTION As customers were leaving a restaurant, they were asked to rate the service they had received. Good service was rated with a 5, fair service with a 3, and poor service with a 1. The tally sheet compiled by the questioner is shown below. What was the restaurant's average score on this survey?

Type of service	Point value	Number
Good	5	卌 卌 卌 卌 卌 卌 卌 卌 卌
Fair	3	卌 卌 卌 |||
Poor	1	卌 ||||

WRITING

97. Explain the error in the student's work shown below.

Evaluate $-a + 3a$ for $a = -6$.

$$\begin{aligned} -a + 3a &= -6 + 3(-6) \\ &= -6 + (-18) \\ &= -24 \end{aligned}$$

98. Explain how we can use a stopwatch to find the distance traveled by a falling object.

99. Write a definition for each of these business words: *revenue, markup,* and *profit.*

100. What is a formula?

101. In this section we *substituted* a number for a variable. List some other uses of the word *substitute* that you encounter in everyday life.

102. What is a **weighted mean**?

103. Show the misunderstanding that occurs if we don't write parentheses around -8 when evaluating the expression $2x + 10$ for $x = -8$.

$$\begin{aligned} 2x + 10 &= 2 - 8 + 10 \\ &= -6 + 10 \\ &= 4 \end{aligned}$$

104. Explain why the following instruction is incomplete.

Evaluate the algebraic expression $3a^2 - 4$.

105. What occupation might use a formula that finds:

a. target heart rate after a workout

b. gas mileage of a car

c. age of a fossil

d. equity in a home

e. dose to administer

f. cost-of-living index

106. A car travels at a rate of 65 mph for 15 minutes. What is wrong with the following thinking?

$$\begin{aligned} d &= rt \\ &= 65(15) \\ &= 975 \end{aligned}$$

The car travels 975 miles in 15 minutes.

SECTION 1.9

Applications Introduction: Perimeter and Area

In Section 1.9, we will study some important concepts from geometry: perimeter and area. The **perimeter** of a plane (two-dimensional, flat) figure is the distance around the figure. Perimeter is measured in American units, such as inches, feet, yards, and in metric units, such as millimeters, meters, and kilometers. The **area** of a plane figure is the amount of surface it encloses. Area is measured in square units, such as square inches, square feet, and square meters.

1. Tell which concept applies to each of the following situations, *perimeter* or *area.*

- **a.** Refinishing a dance floor
- **b.** Installing fencing around a garden
- **c.** Tinting a car's windows
- **d.** Laying a beach towel out on the sand
- **e.** Walking around the sidelines of a soccer field
- **f.** Sewing lace around the edges of a handkerchief
- **g.** Wallpapering the walls of a bedroom

2. Match each item to its approximate area.

a. The state of North Carolina	**i.** 2,400 square feet
b. A $20 bill	**ii.** 1 square inch
c. A UFC octagonal fighting ring	**iii.** 54,000 square miles
d. A piece of notebook paper	**iv.** 8,000 square yards
e. Average U.S. home size	**v.** 94 square inches
f. A postage stamp	**vi.** 750 square feet
g. Grass turf on the University of Michigan football field	**vii.** 16 square inches

SECTION 1.9

Perimeter and Area

Objectives

1. Find perimeters.
2. Find the perimeters of combinations of figures.
3. Find areas.
4. Find the area of combinations of figures.

ARE YOU READY?

The following problems review some of the basic skills that are needed when finding perimeter and area.

Evaluate each expression.

1. Evaluate: $4(86)$

2. Evaluate: $2(18) + 2(33)$

3. Evaluate: 15^2

4. What is $\frac{1}{2}$ of 12?

5. Evaluate: $15(8) - 5^2$

6. a. How many feet are in 1 yard?

b. How many inches are in 1 foot?

c. How many inches are in 1 yard?

In this section, we will discuss how to find perimeters and areas of several types of geometric figures. Perimeter is important when estimating the cost of fencing a yard or installing crown molding in a room. Area is important when calculating the cost of carpeting, painting a room, or fertilizing a lawn.

1 Find perimeters.

To find the **perimeter** of a plane (two-dimensional, flat) geometric figure, such as a square, a rectangle or triangle, we find the distance around the figure by computing the sum of the lengths of its sides. Perimeter is measured in American units, such as inches, feet, yards, and in metric units, such as millimeters, meters, and kilometers.

> ***The Language of Algebra*** When you hear the word **perimeter**, think of the distance around the "rim" of a flat figure.

Triangle

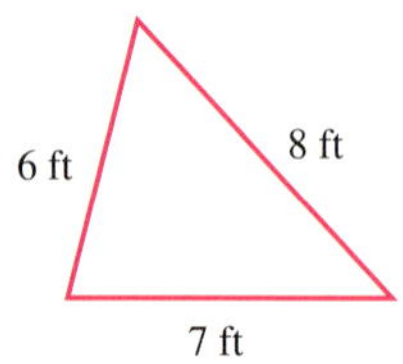

$P = 6 + 7 + 8$
$P = 21$

The perimeter is 21 ft.

Square

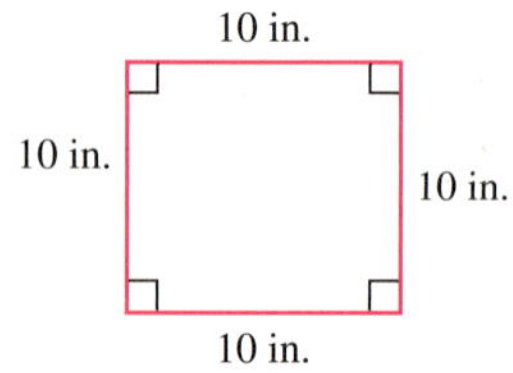

$P = 10 + 10 + 10 + 10$
$P = 40$

The perimeter is 40 in.

Rectangle

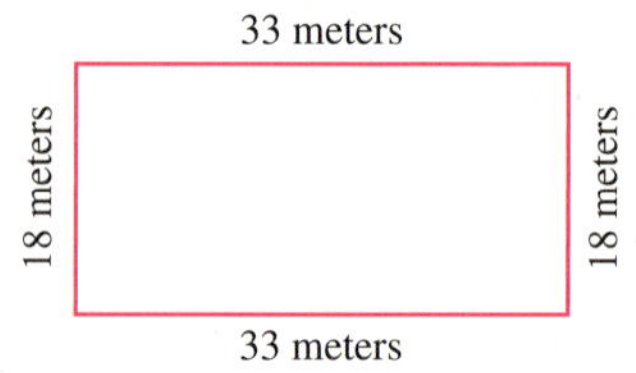

$P = 18 + 33 + 18 + 33$
$P = 102$

The perimeter is 102 meters

For some figures, such as a square and a rectangle, we can simplify the calculations by using a perimeter formula. Since a square has four sides of equal length s, its perimeter P is $s + s + s + s$, or $4s$.

Perimeter of a Square

If a square has a side of length s, its perimeter P is given by the formula

$$P = 4s$$

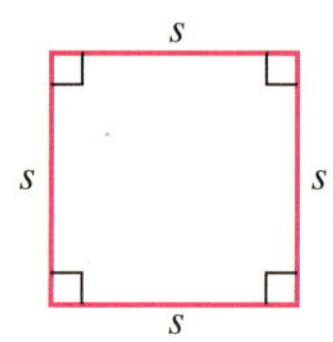

Self Check 1

A Scrabble game board has a square shape with sides of length 38 cm. Find the perimeter of the game board.

Now Try **Problems 17 and 19**

EXAMPLE 1 Find the perimeter of a square whose sides are 7 meters long.

Strategy We will substitute 7 for s in the formula $P = 4s$ and evaluate the right side.

WHY The variable P represents the unknown perimeter of the square.

Solution

$P = 4s$ This is the formula for the perimeter of a square.

$P = 4(7)$ Substitute 7 for s, the length of one side of the square.

$P = 28$ Do the multiplication.

The perimeter of the square is 28 meters.

Since a rectangle has two lengths l and two widths w, its perimeter P is given by $l + w + l + w$, or $2l + 2w$.

Perimeter of a Rectangle

If a rectangle has length l and width w, its perimeter P is given by the formula

$P = 2l + 2w$

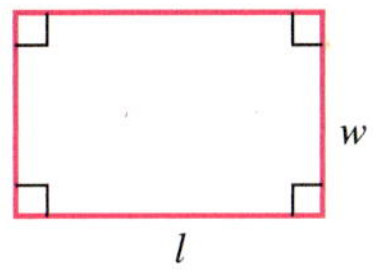

Caution! When finding the perimeter of a polygon, the lengths of the sides must be expressed in the same units.

EXAMPLE 2 Find the perimeter of the rectangle shown on the right, in inches.

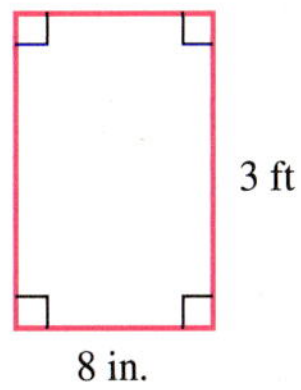

Strategy We will express the width of the rectangle in inches and then use the formula $P = 2l + 2w$ to find the perimeter of the figure.

WHY We can only add quantities that are measured in the same units.

Solution Since 1 foot = 12 inches, we can convert 3 feet to inches by multiplying 3 feet 12. Therefore,

The width of the rectangle is $3 \cdot 12$ inches = 36 inches. We can now substitute 8 for l, the length, and 36 for w, the width, in the formula for the perimeter of a rectangle.

$P = 2l + 2w$ This is the formula for the perimeter of a rectangle.

$P = 2(8) + 2(36)$ Substitute 8 for l, the length, and 36 for w, the width.

$= 16 + 72$ Do the multiplication.

$= 88$ Do the addition.

$$\begin{array}{r} \overset{1}{36} \\ \times 2 \\ \hline 72 \end{array} \qquad \begin{array}{r} 16 \\ +72 \\ \hline 88 \end{array}$$

The perimeter of the rectangle is 88 inches.

Self Check 2

Find the perimeter (in inches) of a rectangle that has a length of 34 inches and a width of 2 feet.

Now Try **Problem 21**

The Language of Mathematics *Triangles* have many real-life applications in such areas as construction, art, and navigation. Two special types of triangles are shown below.

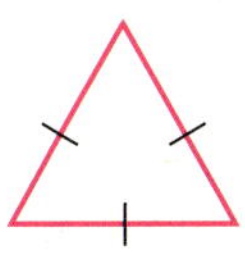

Equilateral triangle
(all sides equal length)

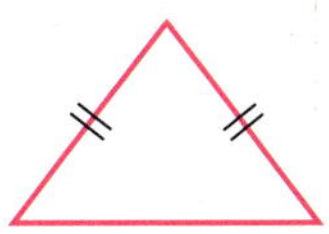

Isosceles triangle
(at least two sides of equal length)

Self Check 3

If one of the sides of equal length of an isosceles triangle is 15 meters long, and its base is 28 meters long. What is the perimeter?

Now Try **Problem 25**

EXAMPLE 3 ***Structural Engineering*** The truss shown below is made up of three parts that form an isosceles triangle. Find the number of linear feet of lumber that was used to make the truss.

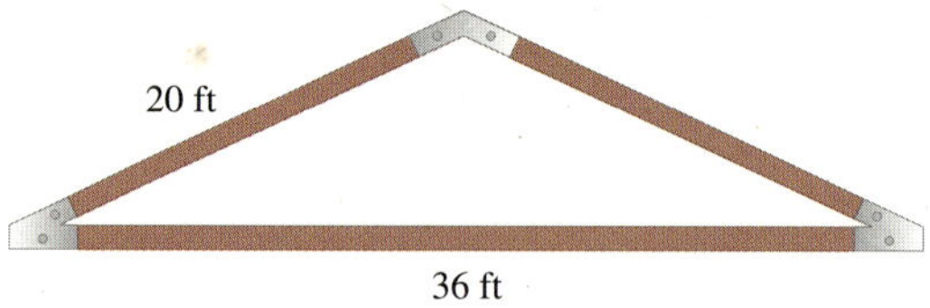

Strategy We will find the perimeter of the triangle.

WHY The number of linear feet of lumber used to make the truss is simply the distance around it. In other words, it is the perimeter.

Solution The sides of the truss form an isosceles triangle with a base that is 36 feet long and one side that is 20 feet long. Since an isosceles triangle has two sides of equal length, the missing side length must also be 20 feet. Now that we know the lengths of the three sides, we can find their sum.

$$P = 20 + 20 + 36$$
$$P = 76$$

The number of linear feet of lumber used to make the truss is 76 feet.

2 Find the perimeter of combinations of figures.

To find the perimeter of the figure shown below, we need to know the values of x and y. Since the figure is a combination of two rectangles, we can use subtraction to find them.

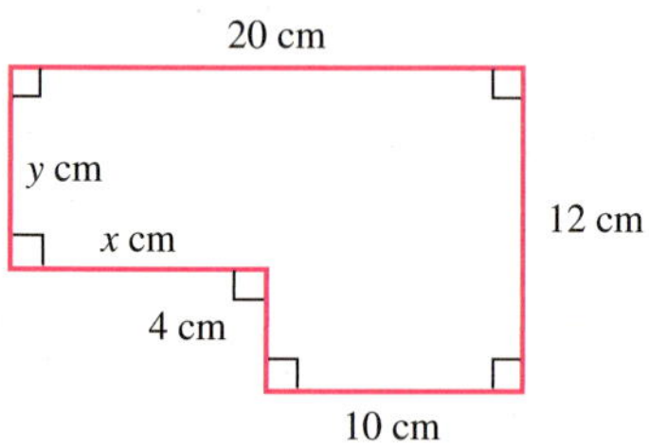

cm stands for centimeters.

$$x = 20 - 10 \qquad \text{and} \qquad y = 12 - 4$$
$$= 10 \text{ cm} \qquad\qquad\qquad = 8 \text{ cm}$$

The perimeter P of the figure is

$$P = 20 + 12 + 10 + 4 + x + y$$
$$P = 20 + 12 + 10 + 4 + 10 + 8$$
$$P = 64$$

The perimeter is 64 centimeters.

3 Find areas.

The **area** of a plane (two-dimensional, flat) geometric figure is the amount of surface that it encloses. Area is measured in square units, such as square inches, square feet, square yards, and square meters (written as in.^{2}, ft^2, yd^2, and m^2, respectively).

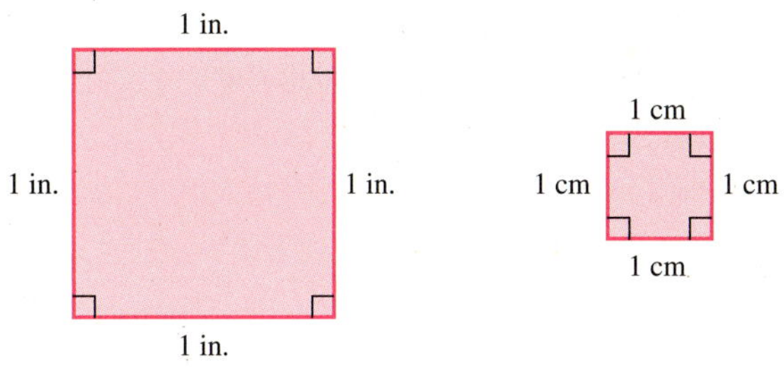

One square inch (1 in.^{2}) **One square centimeter (1 cm^2)**

In everyday life, we often use areas. For example,

- To carpet a room, we buy square yards.
- A can of paint will cover a certain number of square feet.
- To measure vast amounts of land, we often use square miles.
- We buy house roofing by the "square." One square is 100 square feet.

The rectangle shown below has a length of 10 centimeters and a width of 3 centimeters. If we divide the rectangular region into square regions as shown in the figure, each square has an area of 1 square centimeter—a surface enclosed by a square measuring 1 centimeter on each side. Because there are 3 rows with 10 squares in each row, there are 30 squares. Since the rectangle encloses a surface area of 30 squares, its area is 30 square centimeters, which can be written as 30 cm^2.

This example illustrates that to find the area of a rectangle, we multiply its length by its width.

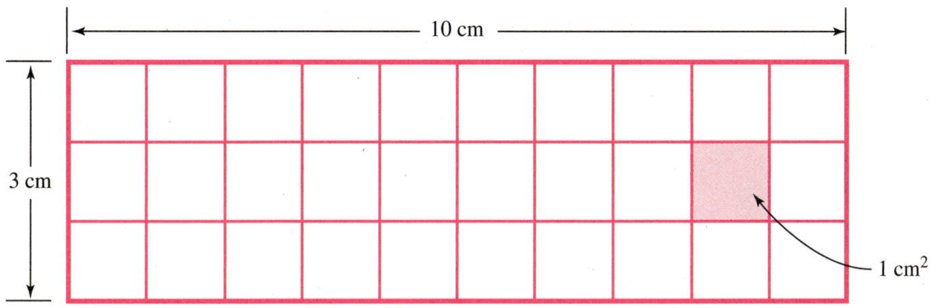

Caution! Do not confuse the concepts of perimeter and area. Perimeter is the distance around a plane geometric figure. It is measured in linear units, such as centimeters, feet, or miles. Area is a measure of the surface enclosed within a plane geometric figure. It is measured in square units, such as square centimeters, square feet, or square miles.

In practice, we do not find areas of geometric figures by counting squares. Instead, we use formulas.

Figure	Name	Formula for Area
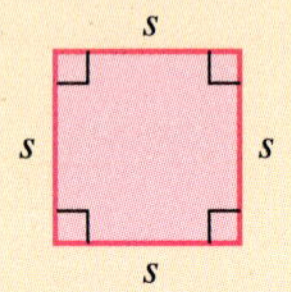	Square	$A = s^2$, where s is the length of one side.
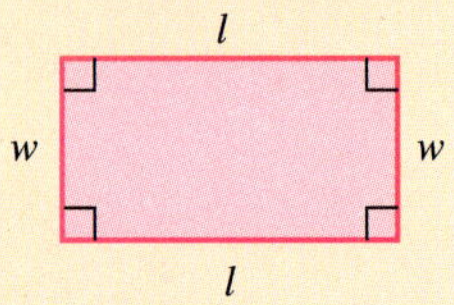	Rectangle	$A = lw$, where l is the length and w is the width.
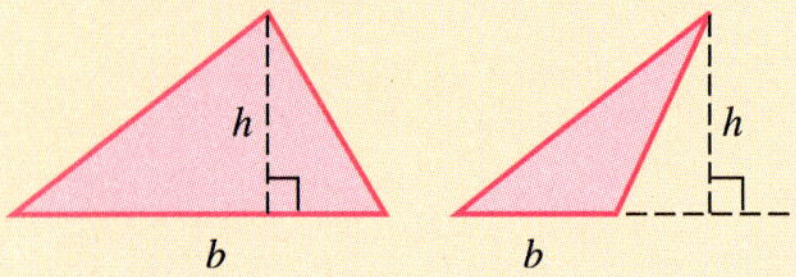	Triangle	$A = \frac{1}{2}bh$, where b is the length of the base and h is the height. The segment perpendicular to the base and representing the height (shown here using a dashed line) is called an **altitude.**

Self Check 4

Find the area of the square shown below.

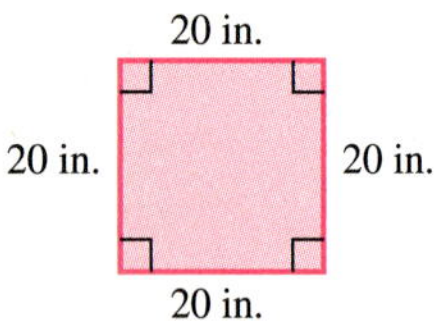

Now Try **Problem 31**

EXAMPLE 4 Find the area of the square shown on the right.

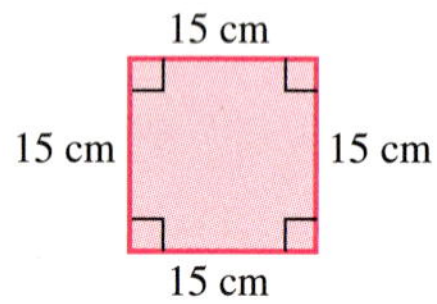

Strategy We will substitute 15 for s in the formula $A = s^2$ and evaluate the right side.

WHY The variable A represents the unknown area of the square.

Solution

$A = s^2$ This is the formula for the area of a square.

$A = 15^2$ Substitute 15 for s, the length of one side of the square.

$A = 225$ Evaluate the exponential expression.

$$\begin{array}{r} 15 \\ \times 15 \\ \hline 75 \\ 150 \\ \hline 225 \end{array}$$

Recall that area is measured in square units. Thus, the area of the square is 225 square centimeters, which can be written as 225 cm^2.

Self Check 5

Find the number of square centimeters in 1 square meter. (Hint: 1 meter = 100 centimeters)

Now Try **Problems 35 and 41**

EXAMPLE 5 Find the number of square feet in 1 square yard.

Strategy A figure is helpful to solve this problem. We will draw a square yard and divide each of its sides into 3 equally long parts.

WHY Since a square yard is a square with each side measuring 1 yard, each side also measures 3 feet.

Solution

$1\ yd^2 = (1\ yd)^2$

$= (3\ ft)^2$ Substitute 3 feet for 1 yard.

$= (3\ ft)(3\ ft)$

$= 9\ ft^2$

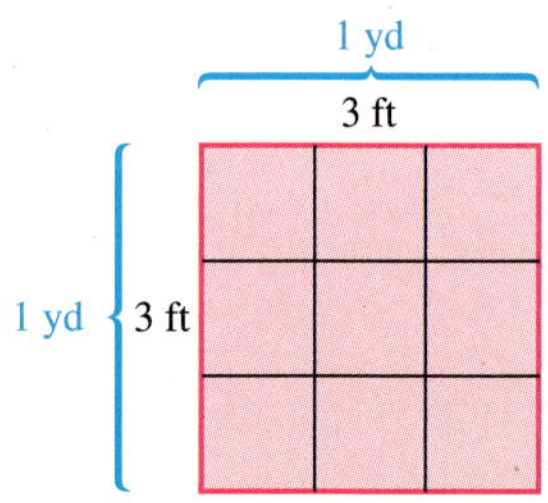

There are 9 square feet in 1 square yard.

EXAMPLE 6 *Women's Sports*

Field hockey is a team sport in which players use sticks to try to hit a ball into their opponents' goal. Find the area of the rectangular field shown on the right. Give the answer in square feet.

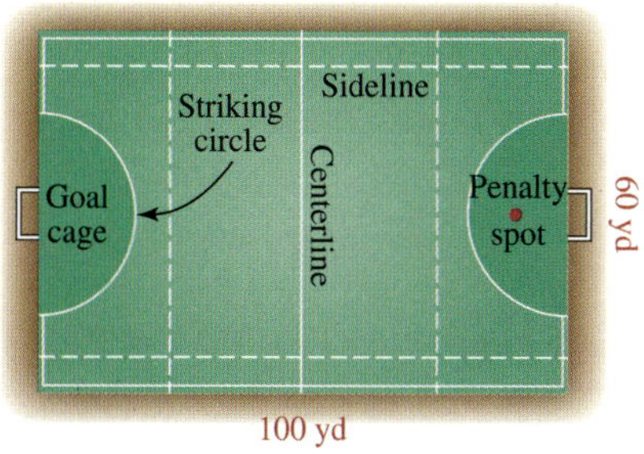

Strategy We will substitute 100 for l and 60 for w in the formula $A = lw$ and evaluate the right side.

WHY The variable A represents the unknown area of the rectangle.

Solution

$A = lw$ — This is the formula for the area of a rectangle.

$A = 100(60)$ — Substitute 100 for l, the length, and 60 for w, the width.

$= 6{,}000$ — Do the multiplication.

The area of the rectangle is 6,000 square yards. Since there are 9 square feet per square yard, we can convert this number to square feet by multiplying 6,000 square yards by 9. Therefore,

The area of the field is $6{,}000 \cdot 9 = 54{,}000 \text{ ft}^2$.

Self Check 6

PING-PONG A regulation-size Ping-Pong table is 9 feet long and 5 feet wide. Find its area in square inches.

Now Try **Problem 43**

THINK IT THROUGH *Dorm Rooms*

"The United States has more than 4,000 colleges and universities, with 2.3 million students living in college dorms."

The New York Times, 2007

The average dormitory room in a residence hall has about 180 square feet of floor space. The rooms are usually furnished with the following items having the given dimensions:

- 2 extra-long twin beds (each is 39 in. wide × 80 in. long × 24 in. high)
- 2 dressers (each is 18 in. wide × 36 in. long × 48 in. high)
- 2 bookcases (each is 12 in. wide × 24 in. long × 40 in. high)
- 2 desks (each is 24 in. wide × 48 in. long × 28 in. high)

How many square feet of floor space are left?

EXAMPLE 7 Find the area of the triangle shown on the right.

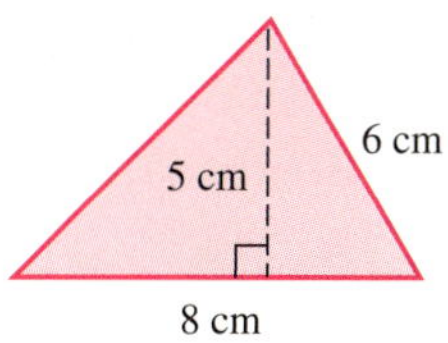

Strategy We will substitute 8 for b and 5 for h in the formula $A = \frac{1}{2}bh$ and evaluate the right side. (The side having length 6 cm is additional information that is not used to find the area.)

WHY The variable A represents the unknown area of the triangle.

Self Check 7

Find the area of the triangle shown below.

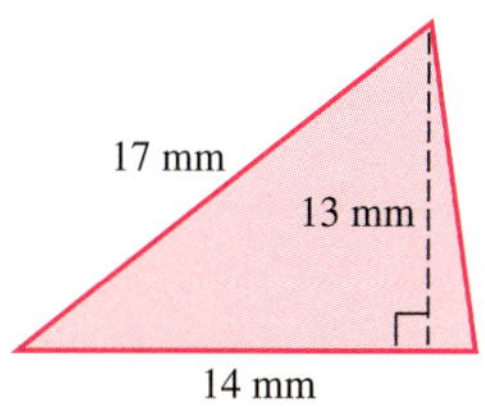

Now Try **Problem 47**

Solution

$$A = \frac{1}{2}bh \quad \text{This is the formula for the area of a triangle.}$$

$$A = \frac{1}{2}(8)(5) \quad \text{Substitute 8 for } b\text{, the length of the base, and 5 for } h\text{, the height.}$$

$$= 4(5) \quad \text{Do the first multiplication: } \tfrac{1}{2}(8) = 4\text{. Think: one-half of 8 is 4.}$$

$$= 20 \quad \text{Complete the multiplication.}$$

The area of the triangle is 20 cm^2.

Self Check 8

Find the area of the triangle shown below.

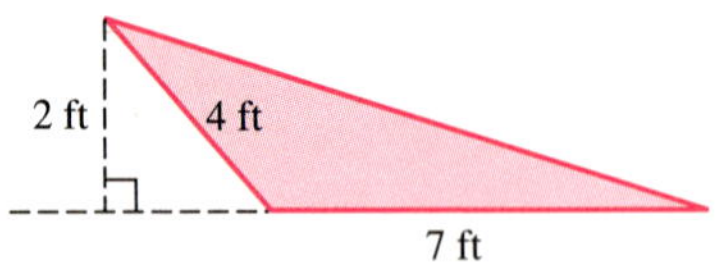

Now Try **Problem 51**

EXAMPLE 8 Find the area of the triangle shown on the right.

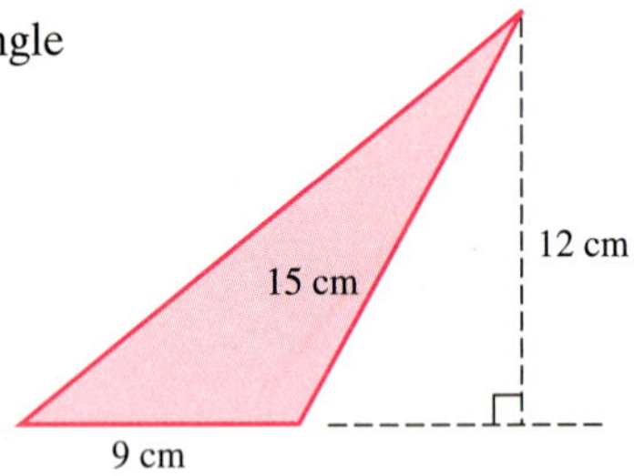

Strategy We will substitute 9 for b and 12 for h in the formula $A = \frac{1}{2}bh$ and evaluate the right side. (The side having length 15 cm is additional information that is not used to find the area.)

WHY The variable A represents the unknown area of the triangle.

Solution In this case, the altitude falls outside the triangle.

$$A = \frac{1}{2}bh \quad \text{This is the formula for the area of a triangle.}$$

$$A = \frac{1}{2}(9)(12) \quad \text{Substitute 9 for } b\text{, the length of the base, and 12 for } h\text{, the height.}$$

$$= \frac{1}{2}(12)(9) \quad \text{Since it's easier to find } \tfrac{1}{2} \text{ of 12 than } \tfrac{1}{2} \text{ of 9, use the commutative property of multiplication to reorder the last two factors.}$$

$$= 6(9) \quad \text{Do the first multiplication: } \tfrac{1}{2}(12) = 6\text{. Think: one-half of 12 is 6.}$$

$$= 54 \quad \text{Do the multiplication.}$$

The area of the triangle is 54 cm^2.

Success Tip The formula for the area of a triangle, $A = \frac{1}{2}bh$, involves a fraction. In DMA 020 Module 2, we will discuss fraction multiplication in more detail. In this module, you will only be asked to multiply $\frac{1}{2}$ and even numbers. You can find the result of such multiplications by simply dividing by 2.

4 Find the area of combinations of figures.

To find the area of an irregular shape, break it up into familiar shapes, such as squares, rectangles, and triangles, find the area of each piece and then add them.

EXAMPLE 9 ***Carpeting a Room*** A living room/dining room has the floor plan shown in the figure. If carpet costs \$29 per square yard, including pad and installation, how much will it cost to carpet both rooms? (Assume no waste.)

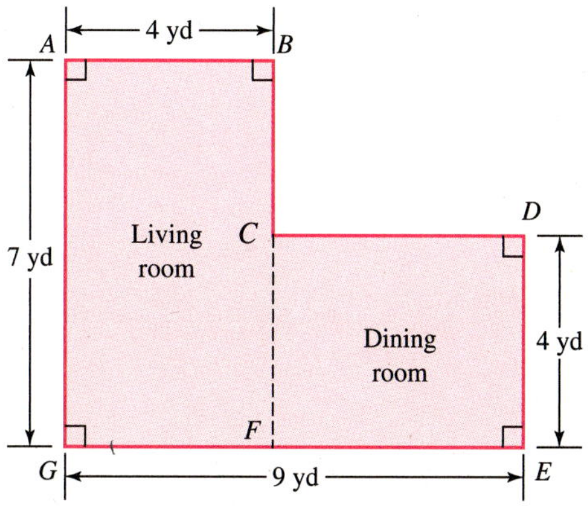

Strategy We will find the number of square yards of carpeting needed and multiply the result by \$29.

WHY Each square yard costs \$29.

Solution First, we must find the total area of the living room and the dining room:

$$A_{\text{total}} = A_{\text{living room}} + A_{\text{dining room}}$$

Since $\overline{CF}$ divides the space into two rectangles, the areas of the living room and the dining room are found by multiplying their respective lengths and widths. Therefore, the area of the living room is $4 \text{ yd} \cdot 7 \text{ yd} = 28 \text{ yd}^2$.

The width of the dining room is given as 4 yd. To find its length, we subtract:

$$m(\overline{CD}) = m(\overline{GE}) - m(\overline{AB}) = 9 \text{ yd} - 4 \text{ yd} = 5 \text{ yd}$$

Thus, the area of the dining room is $5 \text{ yd} \cdot 4 \text{ yd} = 20 \text{ yd}^2$. The total area to be carpeted is the sum of these two areas.

$$A_{\text{total}} = A_{\text{living room}} + A_{\text{dining room}}$$
$$A_{\text{total}} = 28 \text{ yd}^2 + 20 \text{ yd}^2$$
$$= 48 \text{ yd}^2$$

$$\begin{array}{r} 48 \\ \times 29 \\ \hline 432 \\ 960 \\ \hline 1{,}392 \end{array}$$

At \$29 per square yard, the cost to carpet both rooms will be $48 \cdot \$29$, or \$1,392.

Self Check 9

Find the area of the shaded figure.

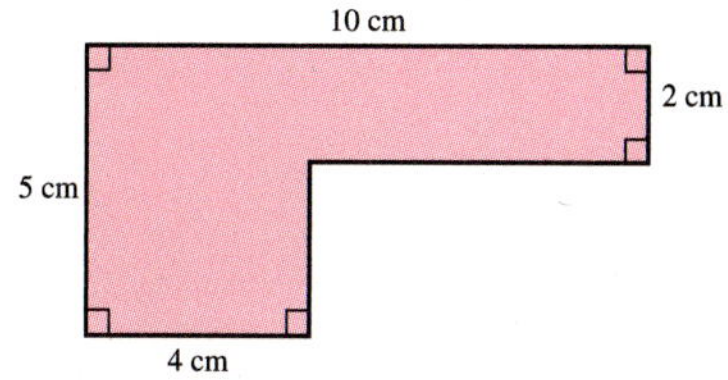

Now Try **Problems 55 and 59**

EXAMPLE 10 Find the area of the shaded region shown on the right.

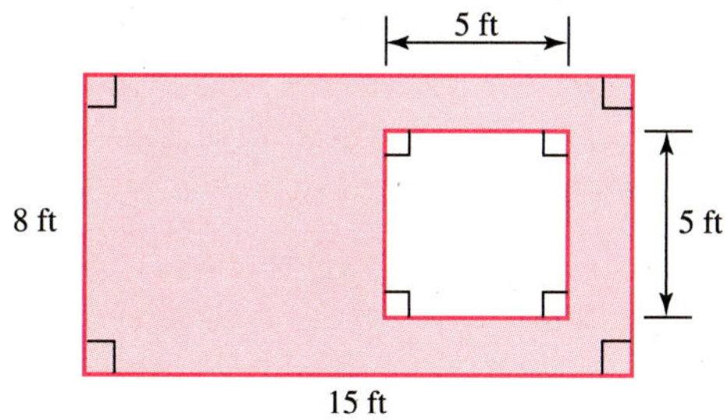

Strategy We will subtract the unwanted area of the square from the area of the rectangle.

Area of shaded region = Area of rectangle − Area of square

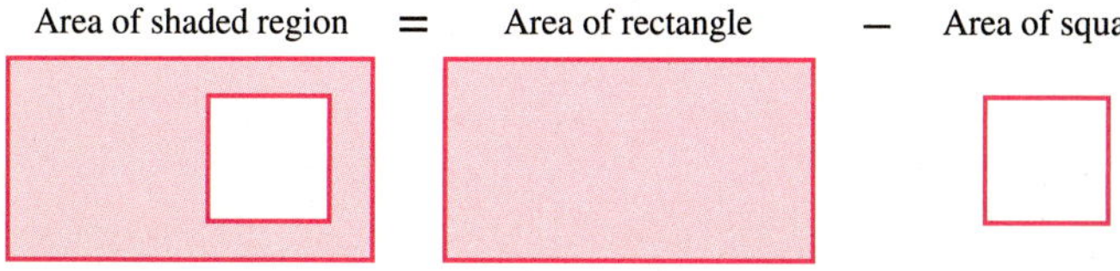

Self Check 10

Find the area of the shaded region shown below.

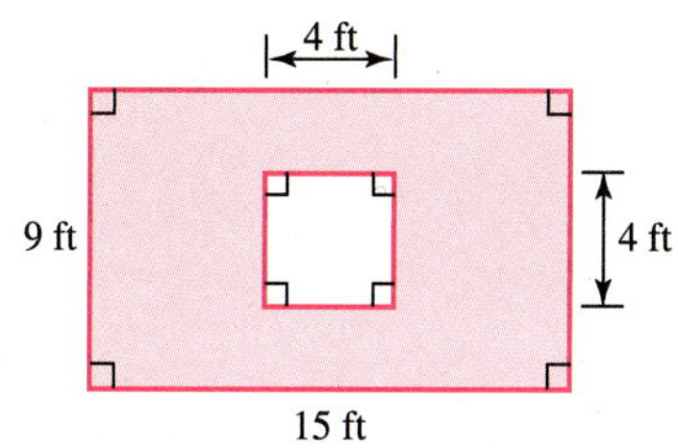

Now Try **Problem 63**

WHY The area of the rectangular-shaped shaded figure does not include the square region inside of it.

Solution

$A_{\text{shaded}} = lw - s^2$ — The formula for the area of a rectangle is $A = lw$. The formula for the area of a square is $A = s^2$.

$A_{\text{shaded}} = 15(8) - 5^2$ — Substitute 15 for the length l and 8 for the width w of the rectangle. Substitute 5 for the length s of a side of the square.

$= 120 - 25$

$= 95$

$$\begin{array}{r} \overset{4}{15} \\ \times 8 \\ \hline 120 \end{array} \qquad \begin{array}{r} 120 \\ -25 \\ \hline 95 \end{array}$$

The area of the shaded region is 95 ft^2.

ANSWERS TO SELF CHECKS

1. 152 cm **2.** 116 in. **3.** 58 m **4.** 400 $in.^2$ **5.** 10,000 cm^2 **6.** 6,480 $in.^2$ **7.** 91 mm^2 **8.** 7 ft^2 **9.** 32 cm^2 **10.** 119 ft^2

SECTION 1.9 STUDY SET

VOCABULARY

Fill in the blanks.

1. The distance around a plane (flat) geometric figure is called its ________.

2. The ________ of a plane (flat) geometric figure is measured in linear units such as inches, feet, and miles.

3. The measure of the surface enclosed by a plane (flat) geometric figure is called its _____.

4. If each side of a square measures 1 foot, the area enclosed by the square is 1 _______ foot.

5. The _____ of a plane (flat) geometric figure is measured in square units.

6. The segment that represents the height of a triangle is called an _______.

CONCEPTS

7. The figure below shows a kitchen floor that is covered with 1-foot-square tiles. Without counting *all* of the squares, determine the area of the floor.

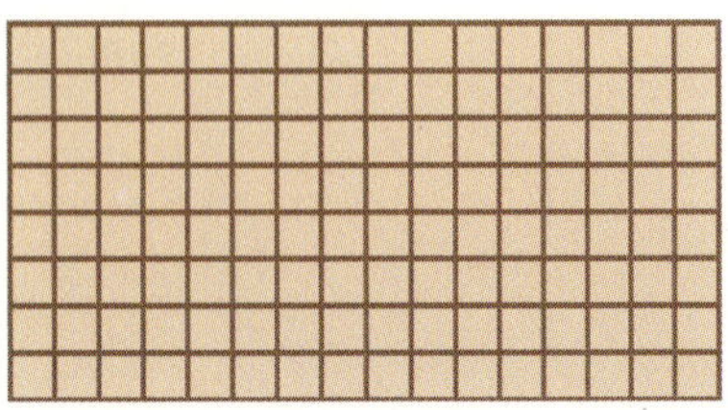

8. Tell which concept applies, perimeter or area.

a. The length of a walk around New York's Central Park

b. The amount of office floor space in the White House

c. The amount of fence needed to enclose a playground

d. The amount of land in Yellowstone National Park

9. Give the formula for the perimeter of a

a. square **b.** rectangle

10. Give the formula for the area of a

a. square **b.** rectangle

c. triangle

11. For each triangle below, draw the altitude to the base b.

a.

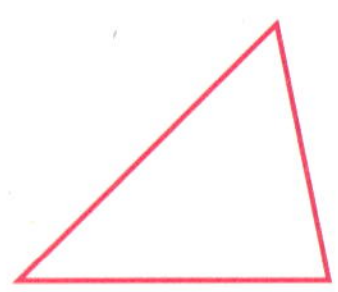

b

b.

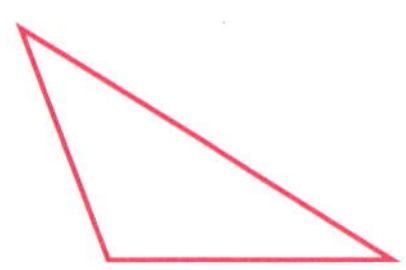

b

12. For each figure below, label the base b for the given altitude.

a.

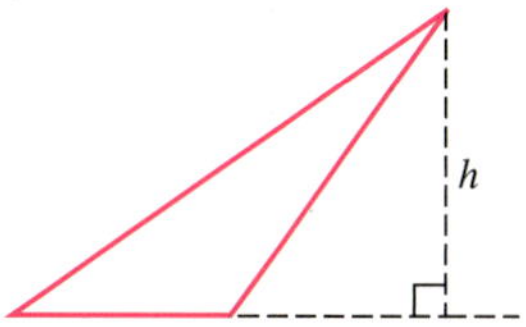

b.

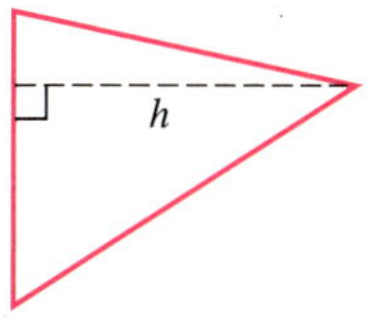

c.

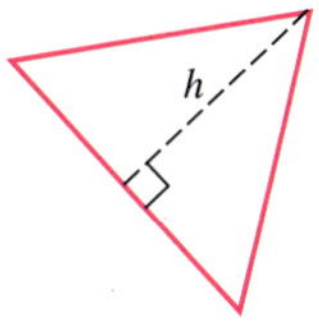

13. The shaded figure below is a combination of what two types of geometric figures?

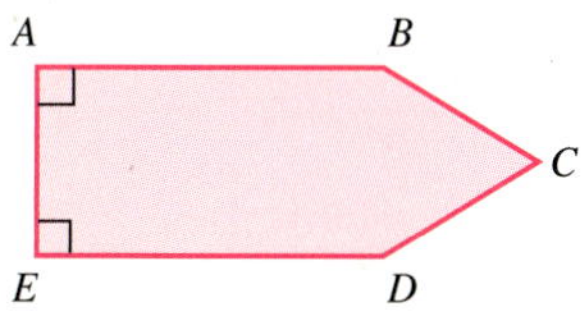

14. Explain how you would find the area of the following shaded figure.

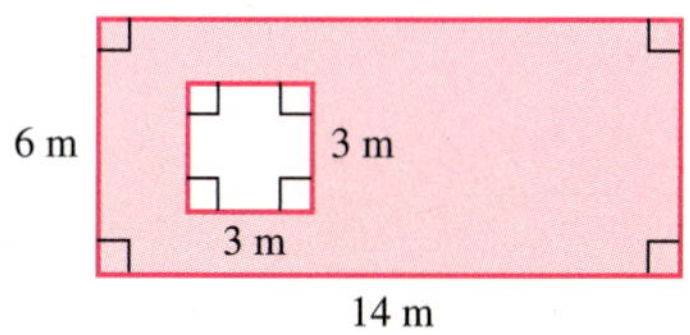

NOTATION

Fill in the blanks.

15. a. The symbol 1 in.^{2} means one _______ _______.

b. One square meter is expressed as _____.

16. In the figure below, the symbol ⌉ indicates that the dashed line segment, called an *altitude,* is ___________ to the base.

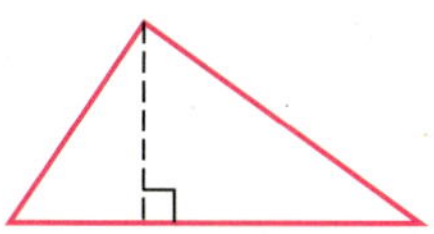

GUIDED PRACTICE

Find the perimeter of each square. See Example 1.

17.

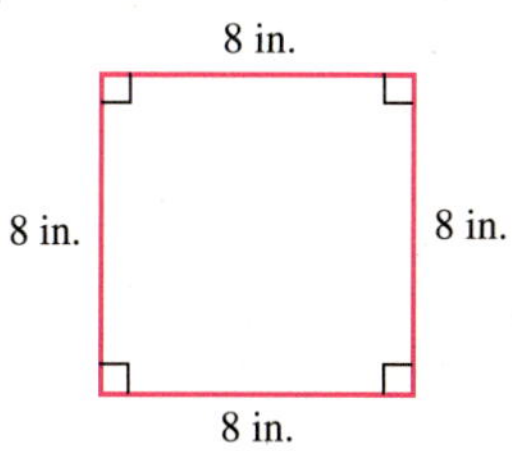

18.

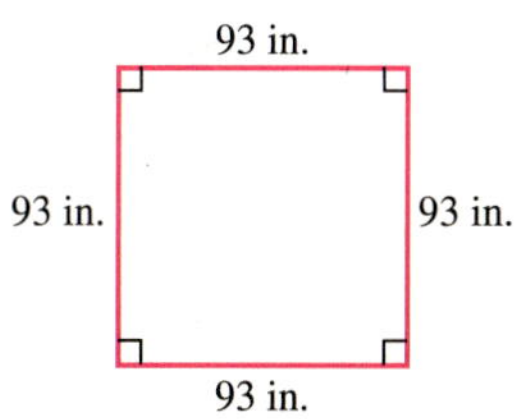

19. A square with sides 6 miles long

20. A square with sides 4 yards long

Find the perimeter of each rectangle, in inches. See Example 2.

21.

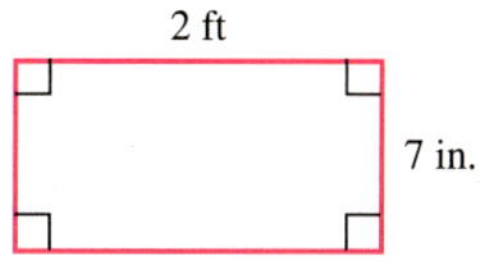

22.

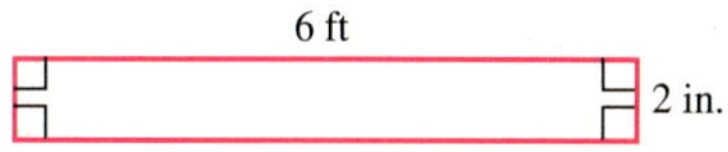

23.

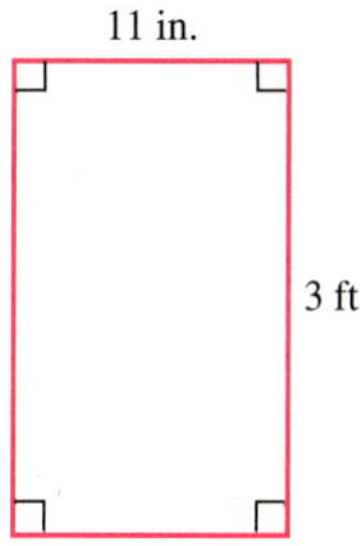

24.

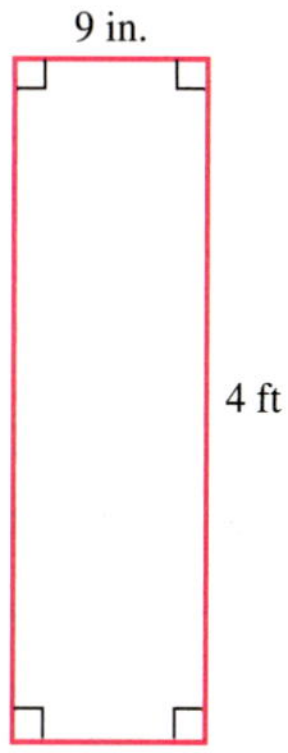

Find the perimeter of each isosceles triangle.
See Example 3.

25. Each of the sides of equal length is 10 feet long. The length of the base is 15 feet.

26. Each of the sides of equal length is 42 feet long. The length of the base is 10 feet.

Find the perimeter of the figure. See Objective 2.

27.

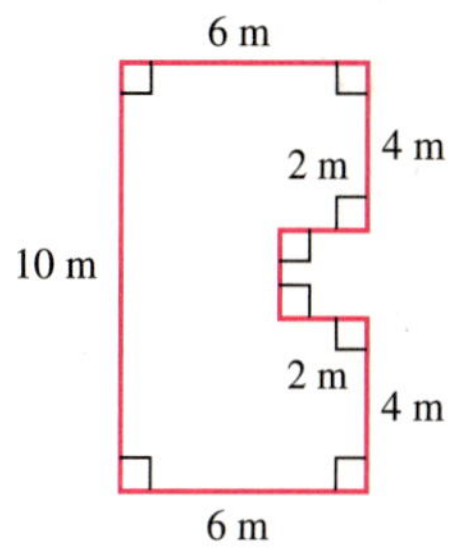

28.

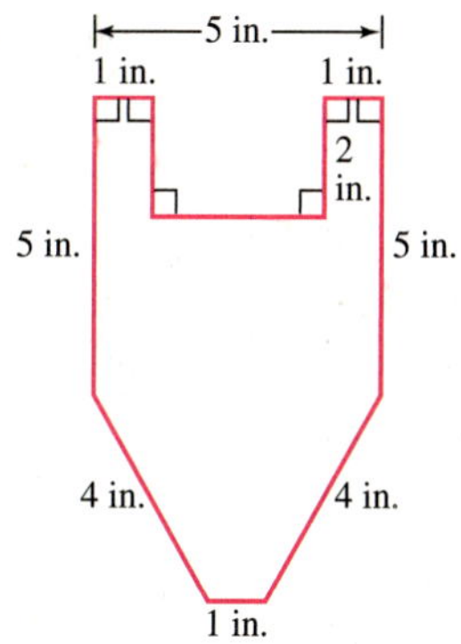

Find x and y. Then find the perimeter of the figure. See Objective 2.

29.

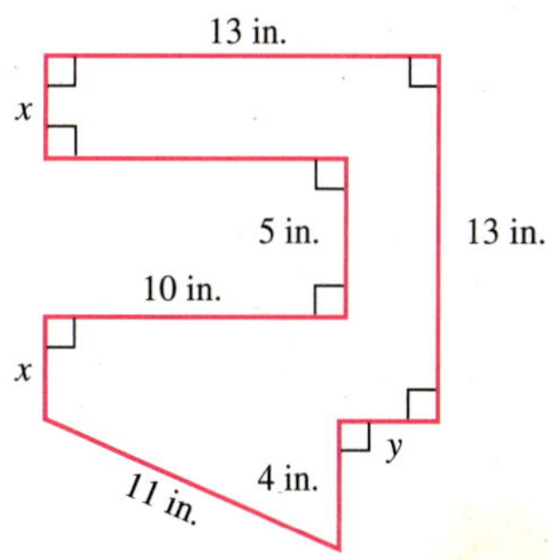

30.

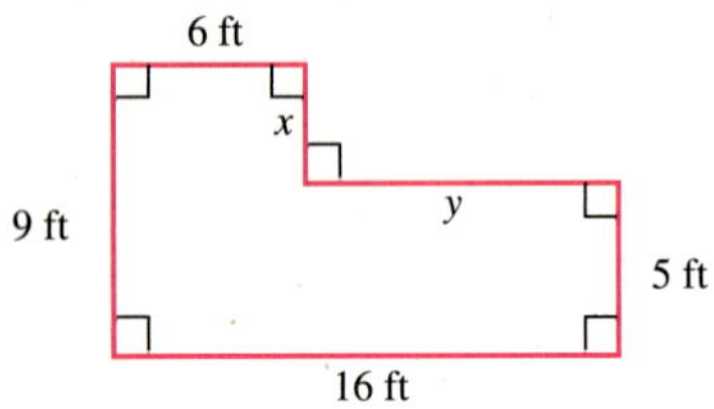

Find the area of each square. **See Example 4.**

31.

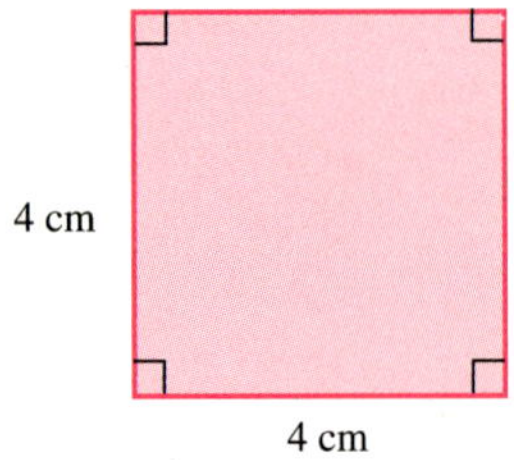

32.

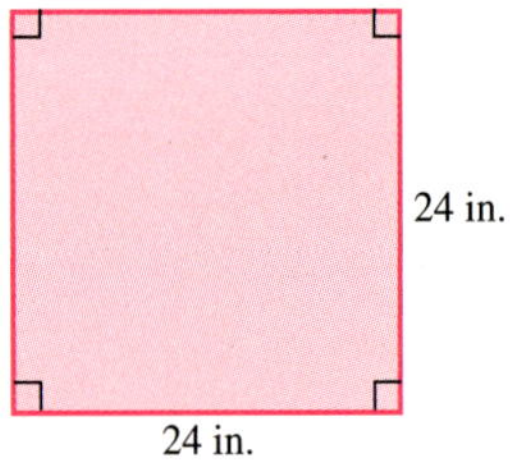

33. A square with sides 25 meters long

34. A square with sides 68 feet long

For Problems 35–42, **see Example 5.**

35. How many square inches are in 1 square foot?

36. How many square inches are in 1 square yard?

37. How many square millimeters are in 1 square meter?

38. How many square decimeters are in 1 square meter? (Hint: 1 meter = 10 decimeters.)

39. How many square feet are in 1 square mile?

40. How many square yards are in 1 square mile?

41. How many square meters are in 1 square kilometer? (Hint: 1 kilometer = 1,000 meters.)

42. How many square dekameters are in 1 square kilometer? (Hint: 100 dekameters = 1 kilometer)

Find the area of each rectangle. Give the answer in square feet. **See Example 6.**

43.

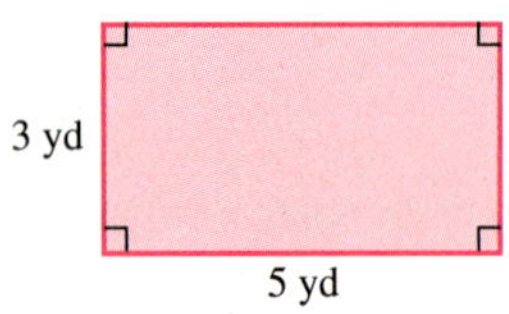

44.

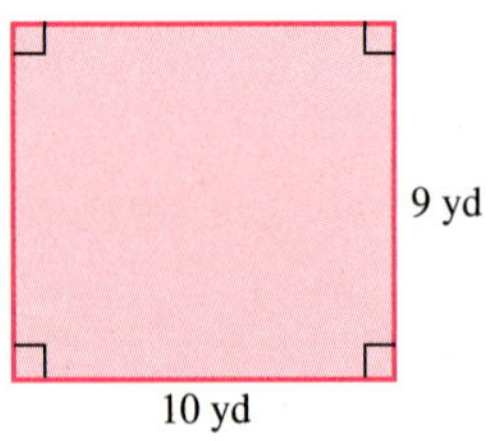

45.

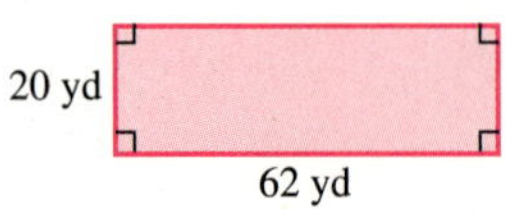

46.

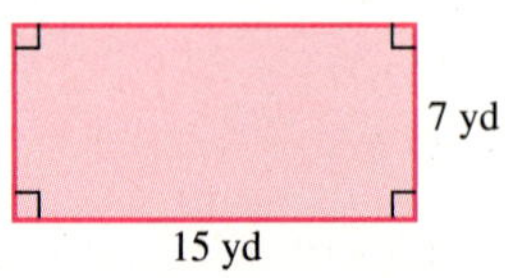

Find the area of each triangle. **See Example 7.**

47.

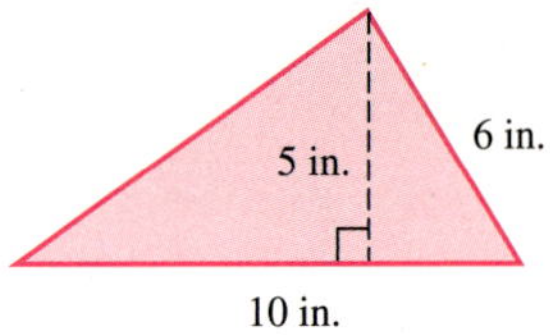

48.

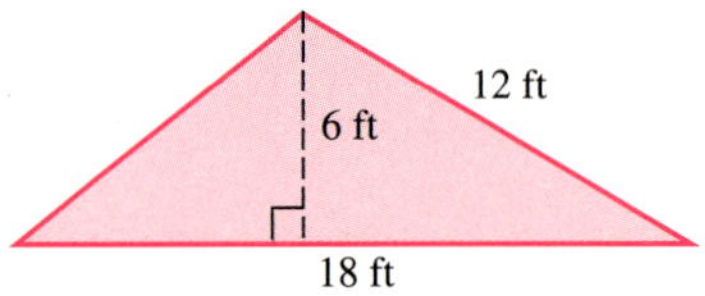

49.

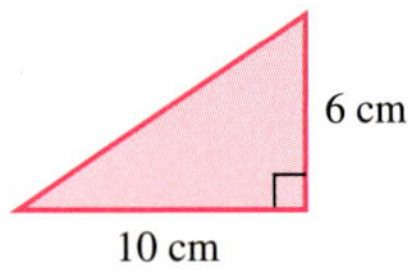

50.

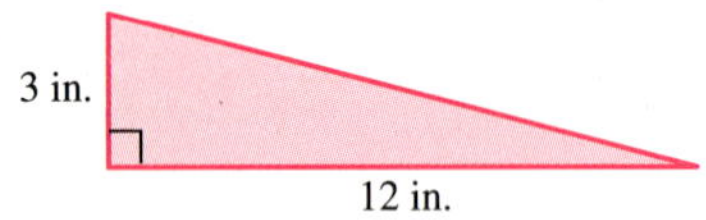

Find the area of each triangle. **See Example 8.**

51.

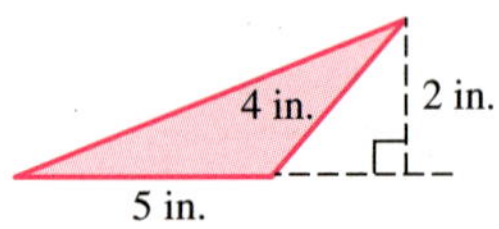

52.

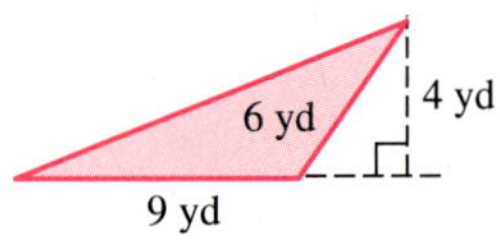

53.

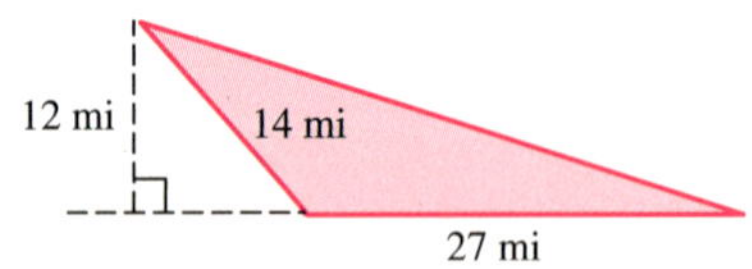

54.

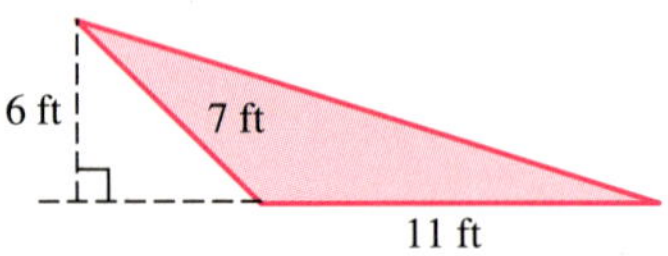

Find the area of each shaded figure. **See Example 9.**

55.

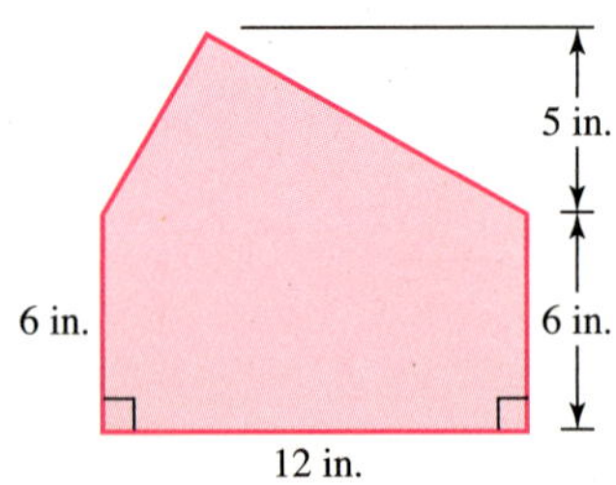

56.

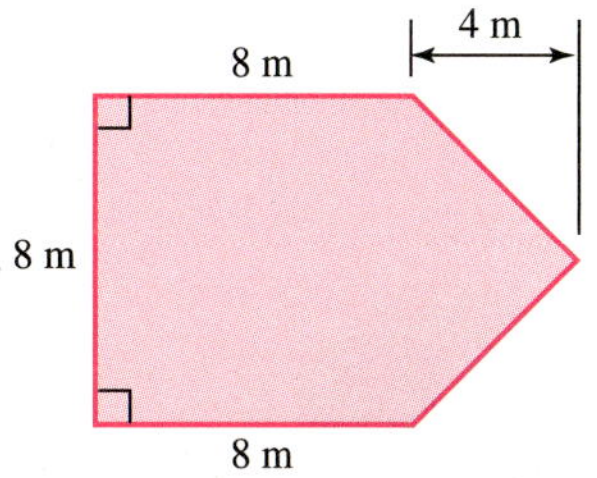

57.

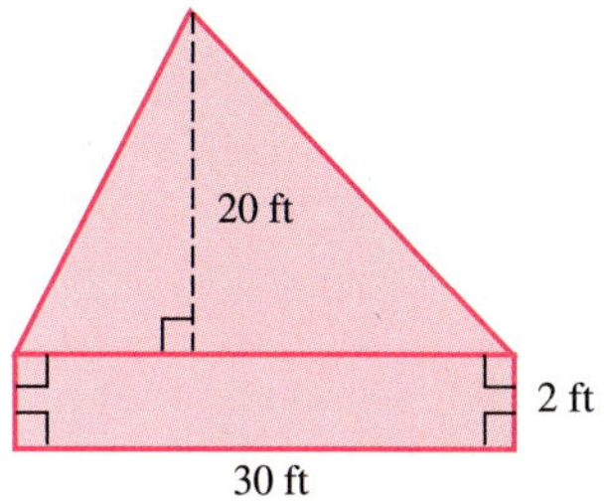

58.

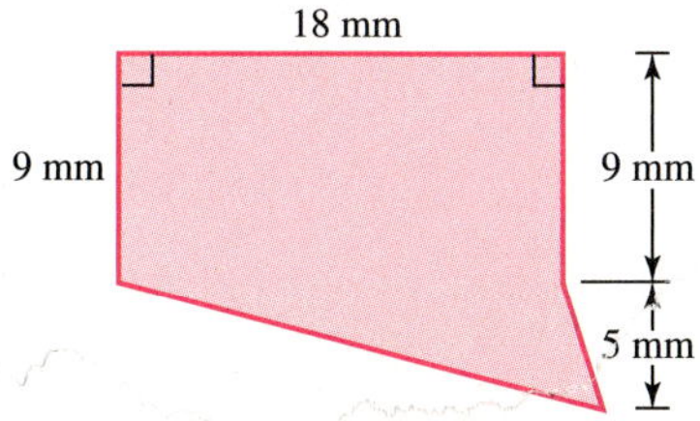

Solve each problem. See Example 9.

59. FLOORING A rectangular family room is 8 yards long and 5 yards wide. At \$30 per square yard, how much will it cost to put down vinyl sheet flooring in the room? (Assume no waste.)

60. CARPETING A rectangular living room measures 10 yards by 6 yards. At \$32 per square yard, how much will it cost to carpet the room? (Assume no waste.)

61. FENCES A man wants to enclose a rectangular yard with fencing that costs \$12 a foot, including installation. Find the cost of enclosing the yard if its dimensions are 110 ft by 85 ft.

62. FRAMES Find the cost of framing a rectangular picture with dimensions of 24 inches by 30 inches if framing material costs \$1 per inch.

Find the area of each shaded figure. See Example.

63.

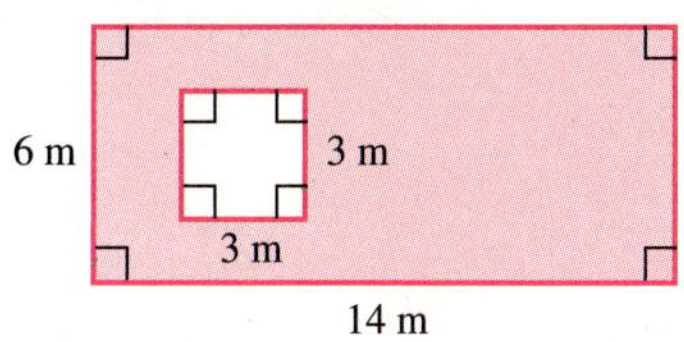

64.

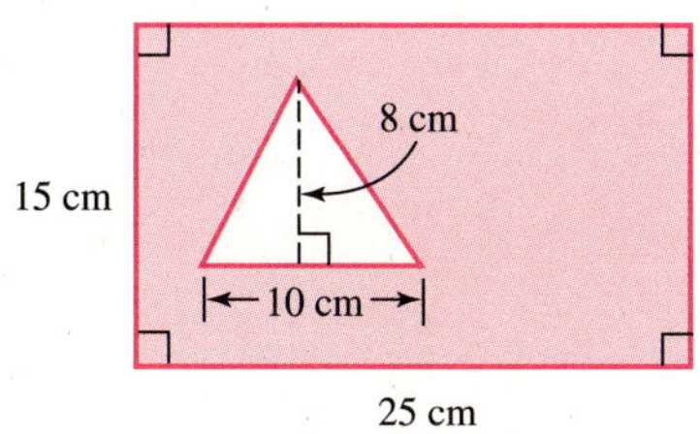

65.

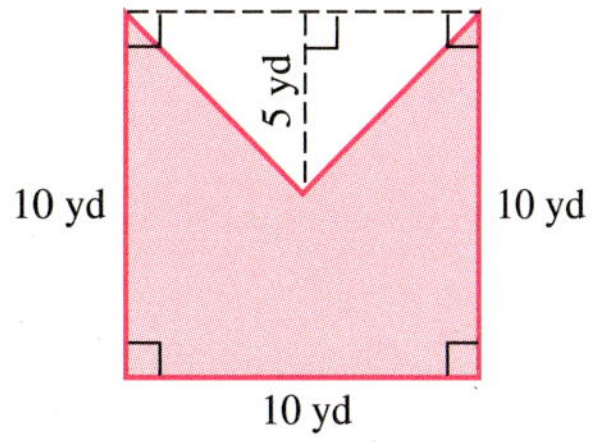

66.

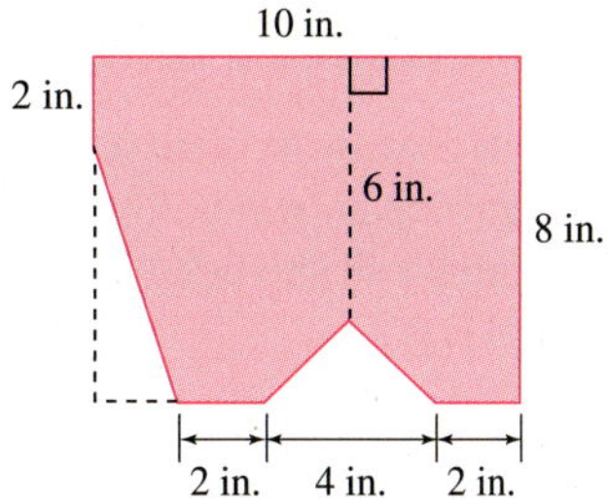

CONCEPT EXTENSIONS

Sketch and label each of the figures.

67. **a.** Two different rectangles, each having a perimeter of 40 in.

b. Find the area of each rectangle in your answer to part a.

68. **a.** Two different rectangles, each having an area of 40 in.^{2}

b. Find the perimeter of each rectangle in your answer to part a.

69. A square with an area of 25 m^2

70. A square with a perimeter of 20 m

71. A triangle with an area of 20 ft^2

72. A figure consisting of a combination of two rectangles, whose total area is 80 ft^2

73. A figure consisting of a combination of a rectangle and a square, whose total area is 164 ft^2

74. Two different isosceles triangles, each having a perimeter of 80 meters, and a length of one side is 22 meters.

75. The area of a rectangle is 36 cm^2, and its length is 3 cm. Find its width.

76. The area of a rectangle is 144 mi^2, and its length is 6 mi. Find its width.

77. The area of a triangle is 54 m^2, and the length of its base is 3 m. Find the height.

78. The area of a triangle is 270 ft^2, and the length of its base is 18 ft. Find the height.

79. The perimeter of a rectangle is 64 mi, and its length is 21 mi. Find its width.

80. The perimeter of a rectangle is 26 yd, and its length is 10 yd. Find its width.

APPLICATIONS

81. GARDENING A gardener wants to plant a border of marigolds around the garden shown below, to keep out rabbits. How many plants will she need if she allows 6 inches between plants?

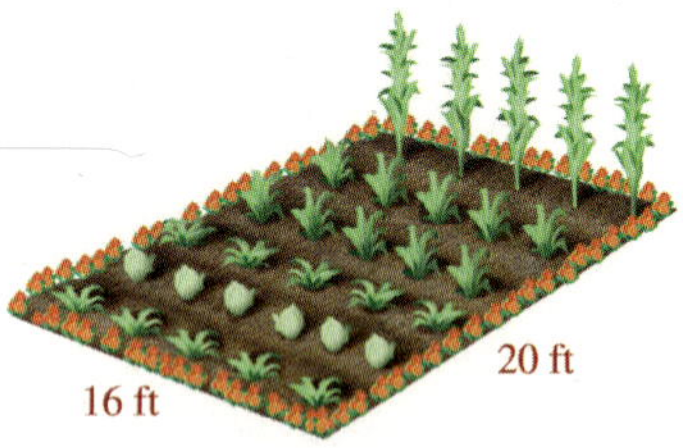

82. COMPARISON SHOPPING Which is more expensive: a ceramic-tile floor costing \$4 per square foot or vinyl costing \$35 per square yard?

83. COMPARISON SHOPPING Which is cheaper: a hardwood floor costing \$4 per square foot or a carpeted floor costing \$32 per square yard?

84. TILES A rectangular basement room measures 14 by 20 feet. Vinyl floor tiles that are 1 ft^2 cost \$3 each. How much will the tile cost to cover the floor? (Assume no waste.)

85. PAINTING The north wall of a barn is a rectangle 23 feet high and 72 feet long. There are five windows in the wall, each 4 by 6 feet. If a gallon of paint will cover 300 ft^2, how many gallons of paint must the painter buy to paint the wall?

86. SAILS If nylon is \$12 per square yard, how much would the fabric cost to make a triangular sail with a base of 12 feet and a height of 24 feet?

87. CARPENTRY How many sheets of 4-foot-by-8-foot sheetrock are needed to drywall the inside walls on the first floor of the barn shown in the next column? (Assume that the carpenters will cover each wall entirely and then cut out areas for the doors and windows.)

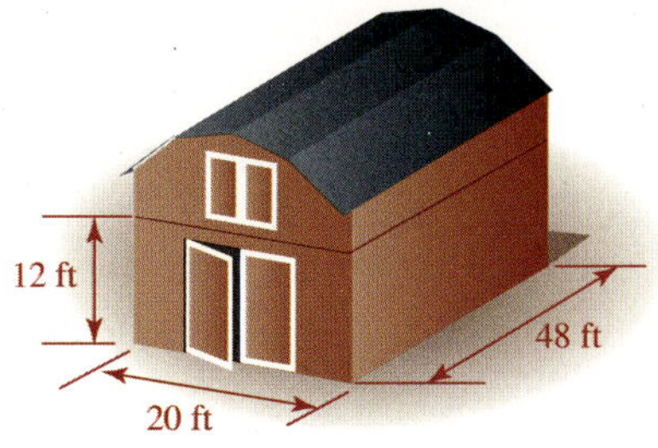

88. CARPENTRY If it costs \$90 per square foot to build a one-story home in northern Wisconsin, find the cost of building the house with the floor plan shown below.

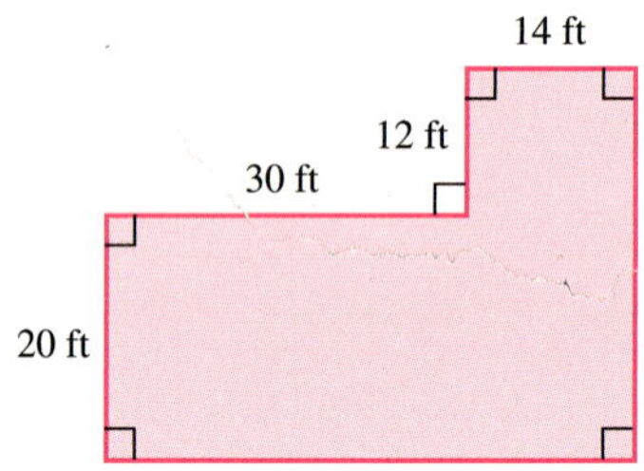

WRITING

89. Explain the difference between perimeter and area.

90. Why is it necessary that area be measured in square units?

91. A student expressed the area of the square in the figure below as 25^2 ft. Explain his error.

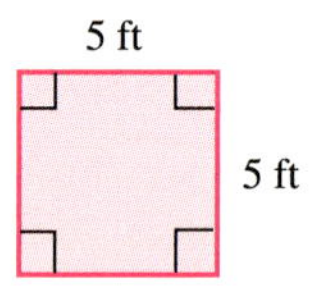

92. Refer to the figure below. What must be done before we can use the formula to find the area of this rectangle?

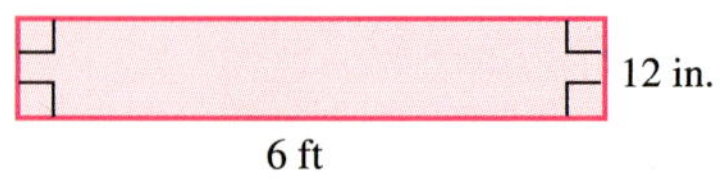

SECTION 1.10
Square Roots

Objectives

1. Find the square root of a perfect square.
2. Evaluate expressions that contain square roots.
3. Evaluate formulas involving square roots.

ARE YOU READY?

The following problems review some basic skills that are needed when working with square roots.

1. Evaluate: 6^2
2. Evaluate: 15^2
3. What positive number, when squared, is 81?
4. What positive number, when squared, is 144?

We have discussed the relationships between addition and subtraction and between multiplication and division. In this section, we explore the relationship between raising a number to a power and finding a root.

1 Find the square root of a perfect square.

When we raise a number to the second power, we are squaring it, or finding its **square.**

The square of 6 is 36, because $6^2 = 36$.
The square of -6 is 36, because $(-6)^2 = 36$.

The **square root** of a given number is a number whose square is the given number. For example, the square roots of 36 are 6 and -6, because either number, when squared, is 36.

Every positive number has two square roots. The number 0 has only one square root. In fact, it is its own square root, because $0^2 = 0$.

Square Root

A number is a **square root** of a second number if the square of the first number equals the second number.

EXAMPLE 1 Find the two square roots of 49.

Strategy We will ask "What positive number and what negative number, when squared, is 49?"

WHY The square root of 49 is a number whose square is 49.

Solution

7 is a square root of 49 because $7^2 = 49$

and

-7 is a square root of 49 because $(-7)^2 = 49$.

Self Check 1

Find the two square roots of 64.

Now Try **Problem 21**

In Example 1, we saw that 49 has two square roots—one positive and one negative. The symbol $\sqrt{}$ is called a **radical symbol** and is used to indicate a positive square root of a nonnegative number. When reading this symbol, we usually drop the word *positive* and simply say *square root.* Since 7 is the positive square root of 49, we can write

$\sqrt{49} = 7$ $\quad$ $\sqrt{49}$ represents the positive number whose square is 49. Read as "the square root of 49 is 7."

When a number, called the **radicand,** is written under a radical symbol, we have a **radical expression.**

Radical symbol ↘ $\sqrt{49}$ ← Radicand

Radical expression

> ***The Language of Algebra*** The symbol $\sqrt{}$ is read as "square root." It also can be read as "radical." For example, $\sqrt{49}$ can be read as "**radical** 49."

Some other examples of radical expressions are:

$$\sqrt{36} \qquad \sqrt{100} \qquad \sqrt{144} \qquad \sqrt{81}$$

To evaluate (or simplify) a radical expression like those shown above, we need to find the positive square root of the radicand. For example, if we evaluate $\sqrt{36}$ (read as "the square root of 36"), the result is

$$\sqrt{36} = 6$$

because $6^2 = 36$.

> ***Caution!*** Remember that the radical symbol asks you to find only the *positive* square root of the radicand. It is incorrect, for example, to say that
>
> ~~$\sqrt{36}$ is 6 and -6~~

The symbol $-\sqrt{}$ is used to indicate the **negative square root** of a positive number. It is the opposite of the positive square root. Since –6 is the negative square root of 36, we can write

$$-\sqrt{36} = -6$$

Read as "the negative square root of 36 is -6" or "the opposite of the square root of 36 is -6." $-\sqrt{36}$ represents the negative number whose square is 36.

If the number under the radical symbol is 0, we have $\sqrt{0} = 0$.

Numbers, such as 36 and 49, that are squares of whole numbers, are called **perfect squares.** To evaluate square root radical expressions, it is helpful to be able to identify perfect square radicands. You need to memorize the following list of perfect squares, shown in red.

Perfect Squares

0 $= 0^2$	**16** $= 4^2$	**64** $= 8^2$	**144** $= 12^2$
1 $= 1^2$	**25** $= 5^2$	**81** $= 9^2$	**169** $= 13^2$
4 $= 2^2$	**36** $= 6^2$	**100** $= 10^2$	**196** $= 14^2$
9 $= 3^2$	**49** $= 7^2$	**121** $= 11^2$	**225** $= 15^2$

A calculator is helpful in finding the square root of a perfect square that is larger than 225.

EXAMPLE 2 Evaluate each square root: **a.** $\sqrt{81}$ **b.** $-\sqrt{100}$

Strategy In each case, we will determine what positive number, when squared, produces the radicand.

WHY The radical symbol $\sqrt{}$ indicates that the positive square root of the number written under it should be found.

Solution

a. $\sqrt{81} = 9$ Ask: What positive number, when squared, is 81? The answer is 9 because $9^2 = 81$.

b. $-\sqrt{100}$ is the opposite (or negative) of the square root of 100. Since $\sqrt{100} = 10$, we have

$$-\sqrt{100} = -10$$

Self Check 2

Evaluate each square root:

a. $\sqrt{144}$

b. $-\sqrt{81}$

Now Try **Problems 25 and 29**

Using Your CALCULATOR **Finding a square root**

We use the $\boxed{\sqrt{}}$ key (square root key) on a scientific calculator to find square roots. For example, to find $\sqrt{729}$, we enter these numbers and press these keys.

729 $\boxed{\sqrt{}}$ $\boxed{27}$

We have found that $\sqrt{729} = 27$. To check this result, we need to square 27. This can be done by entering 27 and pressing the $\boxed{x^2}$ key. We obtain 729. Thus, 27 is the square root of 729.

Some calculator models require keystrokes of $\boxed{\text{2nd}}$ and then $\boxed{\sqrt{}}$ followed by the radicand to find a square root.

2 Evaluate expressions that contain square roots.

In Chapters 1, 2, and 3, we used the order of operations rule to evaluate expressions that involve more than one operation. If an expression contains any square roots, they are to be evaluated at the same stage in your solution as exponential expressions. (See step 2 in the familiar order of operations rule below.)

Order of Operations

1. Perform all calculations within parentheses and other grouping symbols following the order listed in Steps 2–4 below, working from the innermost pair of grouping symbols to the outermost pair.
2. Evaluate all exponential expressions and **square roots.**
3. Perform all multiplications and divisions as they occur from left to right.
4. Perform all additions and subtractions as they occur from left to right.

EXAMPLE 3 Evaluate: **a.** $\sqrt{64} + \sqrt{9}$ **b.** $-\sqrt{25} - \sqrt{225}$

Strategy We will scan the expression to determine what operations need to be performed. Then we will perform those operations, one-at-a-time, following the order of operations rule.

WHY If we don't follow the correct order of operations, the expression can have more than one value.

Self Check 3

Evaluate:

a. $\sqrt{121} + \sqrt{1}$

b. $-\sqrt{9} - \sqrt{196}$

Now Try **Problems 33 and 37**

Solution Since the expression does not contain any parentheses, we begin with step 2 of the rules for the order of operations: Evaluate all exponential expressions and any square roots.

a. $\sqrt{64} + \sqrt{9} = 8 + 3$ Evaluate each square root first.

$= 11$ Do the addition.

b. $-\sqrt{25} - \sqrt{225} = -5 - 15$ Evaluate each square root first.

$= -20$ Do the subtraction.

Self Check 4

Evaluate:

a. $8\sqrt{121}$

b. $-6\sqrt{25} + 2\sqrt{36}$

Now Try **Problems 41 and 45**

EXAMPLE 4 Evaluate: **a.** $6\sqrt{100}$ **b.** $-5\sqrt{16} + 3\sqrt{9}$

Strategy We will scan the expression to determine what operations need to be performed. Then we will perform those operations, one-at-a-time, following the order of operations rule.

WHY If we don't follow the correct order of operations, the expression can have more than one value.

Solution Since the expression does not contain any parentheses, we begin with step 2 of the rules for the order of operations: Evaluate all exponential expressions and any square roots.

a. We note that $6\sqrt{100}$ means $6 \cdot \sqrt{100}$.

$6\sqrt{100} = 6(10)$ Evaluate the square root first.

$= 60$ Do the multiplication.

b. $-5\sqrt{16} + 3\sqrt{9} = -5(4) + 3(3)$ Evaluate each square root first.

$= -20 + 9$ Do the multiplication.

$= -11$ Do the addition.

Self Check 5

Evaluate:

$10 - 4\left[2^2 - (3 + 2)\sqrt{4}\right]$

Now Try **Problems 49 and 53**

EXAMPLE 5 Evaluate: $12 + 3\left[3^2 - (4 - 1)\sqrt{36}\right]$

Strategy We will work within the parentheses first and then within the brackets. Within each set of grouping symbols, we will follow the order of operations rule.

WHY By the order of operations rule, we must work from the *innermost* pair of grouping symbols to the *outermost*.

Solution

$12 + 3\left[3^2 - (4 - 1)\sqrt{36}\right] = 12 + 3\left[3^2 - 3\sqrt{36}\right]$ Do the subtraction within the parentheses.

$= 12 + 3[9 - 3(6)]$ Within the brackets, evaluate the exponential expression and the square root.

$= 12 + 3[9 - 18]$ Do the multiplication within the brackets.

$= 12 + 3[-9]$ Do the subtraction within the brackets.

$= 12 + (-27)$ Do the multiplication.

$= -15$ Do the addition.

3 Evaluate formulas involving square roots.

To evaluate formulas that involve square roots, we replace the letters on one side of the formula with specific numbers and the then use the order of operations rule.

EXAMPLE 6 Evaluate $c = \sqrt{a^2 + b^2}$ for $a = 3$ and $b = 4$.

Strategy In the given formula, we will replace the letter on one side of the formula a with 3 and b with 4. Then we will use the order of operations rule to find the value of the radicand.

WHY We need to know the value of the radicand before we can find its square root.

Solution

$c = \sqrt{a^2 + b^2}$ — This is the formula to evaluate.

$= \sqrt{3^2 + 4^2}$ — Replace a with 3 and b with 4.

$= \sqrt{9 + 16}$ — Evaluate the exponential expressions.

$= \sqrt{25}$ — Do the addition.

$= 5$ — Evaluate the square root.

Self Check 6

Evaluate $a = \sqrt{c^2 - b^2}$ for $c = 17$ and $b = 15$.

Now Try **Problem 67**

ANSWERS TO SELF CHECKS

1. 8 and −8 **2. a.** 12 **b.** −9 **3. a.** 12 **b.** −17 **4. a.** 88 **b.** −18 **5.** 34 **6.** 8

SECTION 1.10 STUDY SET

VOCABULARY

Fill in the blanks.

1. When we raise a number to the second power, we are squaring it, or finding its ________.
2. The square ______ of a given number is a number whose square is the given number.
3. The symbol $\sqrt{\ }$ is called a ________ symbol.
4. Label the *radicand,* the *radical expression,* and the *radical symbol* in the illustration below.

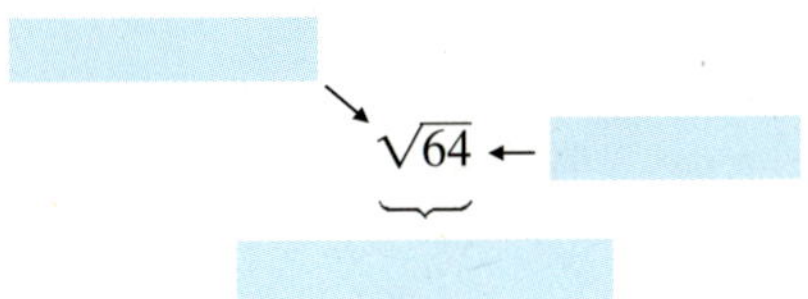

5. Whole numbers such as 36 and 49, that are squares of whole numbers, are called ________ squares.
6. To find the positive square root of 36, we ask "What positive number, when ________, is equal to 36?"

CONCEPTS

Fill in the blanks.

7. The square of 5 is ____, because $5^2 =$ ____.
8. Complete the list of perfect squares: 1, 4, ____, 16, ____, 36, 49, 64, ____, 100, ____, 144, ____, 196, ____.
9. **a.** $\sqrt{49} = 7$, because $\underline{\quad}^2 = 49$.
   b. $\sqrt{4} = 2$, because $\underline{\quad}^2 = 4$.
10. Every positive number has _____ square roots, one positive and one negative. The positive square root of 25 is ____ and the negative square root of 25 is ____. In symbols, we write
    $\sqrt{25} =$ ____ and $-\sqrt{25} =$ ____
11. Evaluate each square root.
    a. $\sqrt{1}$ **b.** $\sqrt{0}$
12. Evaluate each square root.
    a. $\sqrt{121}$ **b.** $\sqrt{144}$ **c.** $\sqrt{169}$
    d. $\sqrt{196}$ **e.** $\sqrt{225}$
13. In what step of the order of operations rule are square roots to be evaluated?
14. Graph $\sqrt{9}$ and $-\sqrt{4}$ on a number line.

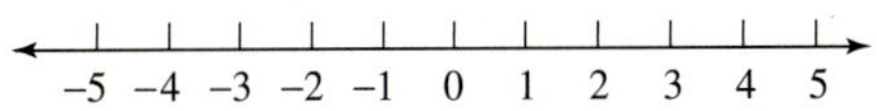

NOTATION

Fill in the blanks.

15. We read $\sqrt{64}$ as "the ________ root of 64" or as "________ 64."
16. The grouping symbols ( ) are ____________ and the symbols [ ] are ________.

17. a. The symbol $\sqrt{}$ is used to indicate a positive ______ _____.

b. The symbol $-\sqrt{}$ is used to indicate the ______ square root of a positive number.

18. $4\sqrt{9}$ means 4 ____ $\sqrt{9}$.

Complete the steps to evaluate the expression.

19. $-\sqrt{49} + \sqrt{64} =$ ____ + ____
$= 1$

20. $2\sqrt{100} - 5\sqrt{25} = 2($ ____ $) - 5($ ____ $)$
$=$ ____ $- 25$
$= -5$

GUIDED PRACTICE

Find the two square roots of each number. See Example 1.

21. 25

22. 1

23. 16

24. 144

Evaluate each square root without using a calculator.
See Example 2.

25. $\sqrt{16}$

26. $\sqrt{64}$

27. $\sqrt{9}$

28. $\sqrt{16}$

29. $-\sqrt{144}$

30. $-\sqrt{121}$

31. $-\sqrt{49}$

32. $-\sqrt{81}$

Evaluate each expression without using a calculator.
See Example 3.

33. $\sqrt{36} + \sqrt{1}$

34. $\sqrt{100} + \sqrt{16}$

35. $\sqrt{81} + \sqrt{49}$

36. $\sqrt{4} + \sqrt{36}$

37. $-\sqrt{144} - \sqrt{16}$

38. $-\sqrt{1} - \sqrt{196}$

39. $-\sqrt{225} + \sqrt{144}$

40. $-\sqrt{169} + \sqrt{16}$

Evaluate each expression without using a calculator.
See Example 4.

41. $4\sqrt{25}$

42. $2\sqrt{81}$

43. $-10\sqrt{196}$

44. $-40\sqrt{4}$

45. $-4\sqrt{169} + 2\sqrt{4}$

46. $-6\sqrt{81} + 5\sqrt{1}$

47. $-8\sqrt{16} + 5\sqrt{225}$

48. $-3\sqrt{169} + 2\sqrt{225}$

Evaluate each expression without using a calculator.
See Example 5.

49. $15 + 4\left[5^2 - (6 - 1)\sqrt{4}\right]$

50. $18 + 2\left[4^2 - (7 - 3)\sqrt{9}\right]$

51. $50 - \left[(6^2 - 24) + 9\sqrt{25}\right]$

52. $40 - \left[(7^2 - 40) + 7\sqrt{64}\right]$

53. $\sqrt{196} + 3\left(5^2 - 2\sqrt{225}\right)$

54. $\sqrt{169} + 2\left(7^2 - 3\sqrt{144}\right)$

55. $\dfrac{\sqrt{16} - 6(2^2)}{\sqrt{4}}$

56. $\dfrac{\sqrt{49} - 3(1^6)}{\sqrt{16} - \sqrt{64}}$

57. $\left(-3 - \sqrt{25}\right)^2$

58. $\left(-1 - \sqrt{144}\right)^2$

59. $5\left(-\sqrt{49}\right)(-2)^2$

60. $\left(-\sqrt{64}\right)(-2)(3)^3$

61. $|\sqrt{49} - 8(4 - 7)|$

62. Evaluate: $\sqrt{(5 - 2)^2 + (8 - 4)^2}$

63. $\dfrac{(7 - 6)^4 + 32}{36 - (\sqrt{16} + 1)^2}$

64. $\dfrac{18|-4-2(3 - 1)|}{-3(\sqrt{9})(-2)}$

Evaluate each formula without using a calculator.
See Example 6.

65. Evaluate $c = \sqrt{a^2 + b^2}$ for $a = 9$ and $b = 12$.

66. Evaluate $c = \sqrt{a^2 + b^2}$ for $a = 6$ and $b = 8$.

67. Evaluate $a = \sqrt{c^2 - b^2}$ for $c = 25$ and $b = 24$.

68. Evaluate $b = \sqrt{c^2 - a^2}$ for $c = 17$ and $a = 8$.

69. $\dfrac{-b + \sqrt{b^2 - 4ac}}{2a}$ for $a = 1$, $b = 2$, $c = -3$

70. $\dfrac{|Ax + By + C|}{\sqrt{A^2 + B^2}}$ for $A = 3$, $B = 4$, $C = -7$, $x = 2$, $y = -1$

71. $\sqrt{(x_2 - x_1)^2 + (y_2 - y_1)^2}$ for $x_1 = -2$, $x_2 = 4$, $y_1 = 4$, $y_2 = -4$

72. $\sqrt{(-1)\left(\dfrac{x + y}{-x - z}\right)}$ for $x = 19$, $y = 17$, and $z = -18$

APPLICATIONS

In the following problems, some lengths are expressed as square roots. Solve each problem by evaluating any square roots. You may need to use a calculator.

73. CARPENTRY Find the length of the slanted side of each roof truss shown below.

a.

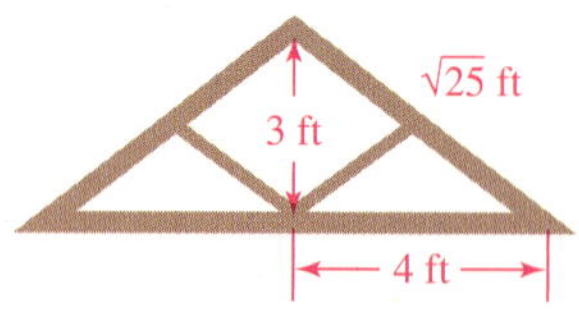

b.

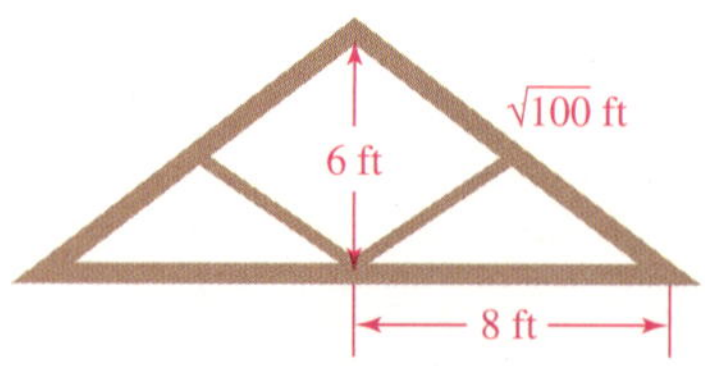

74. RADIO ANTENNAS Refer to the illustration on the next page. How far from the base of the antenna is each guy wire anchored to the ground? (The measurements are in feet.)

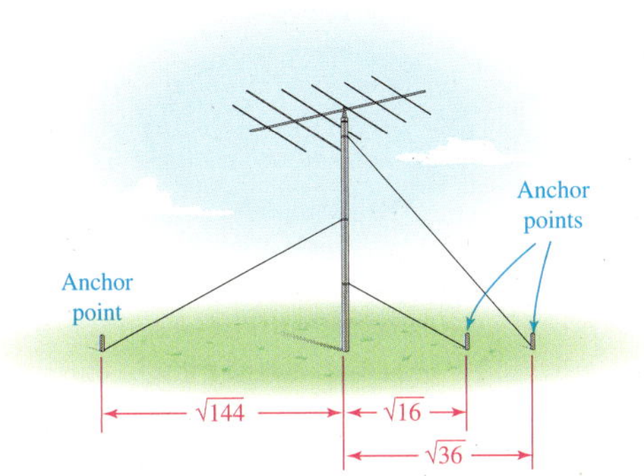

75. CROSSWORD PUZZLES If the area of a square is A square units, then the length of a side of the square is $\sqrt{A}$ units. The world's largest crossword puzzle can be purchased from the *Hammacher Schlemmer Gift Catalog* for \$29.95. The crossword puzzle has a square shape and can be hung on a wall. If it covers 7,056 square inches of wall space, find the length of one of its sides. Answer in inches and also in feet.

76. SURVEYING Refer to the illustration below. Use the imaginary triangles set up by a surveyor to find the length of each lake. (The measurements are in meters.)

a.

b.

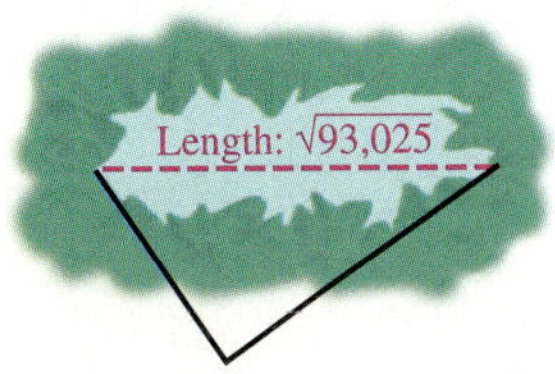

77. FLATSCREEN TELEVISIONS The picture screen on a television set is measured diagonally. What size screen is shown below?

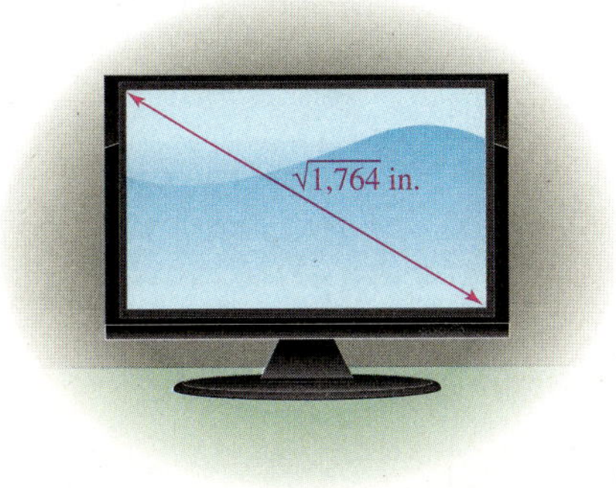

78. LADDERS A painter's ladder is shown in the next column. How long are the legs of the ladder?

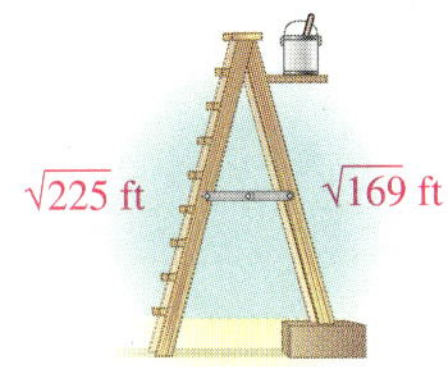

79. SHOELACES The formula
$S = 2\left[H + L + (p - 1)\sqrt{H^2 + V^2}\right]$
can be used to calculate the correct shoelace length for the criss-cross lacing pattern shown in the illustration, where p represents the number of *pairs* of eyelets. Find the correct shoelace length if H(horizontal distance) = 40 mm, L (length of end) = 250 mm, and V(vertical distance) = 30 mm. mm stands for millimeters. (*Source:* Ian's Shoelace Site at www.fieggen.com)

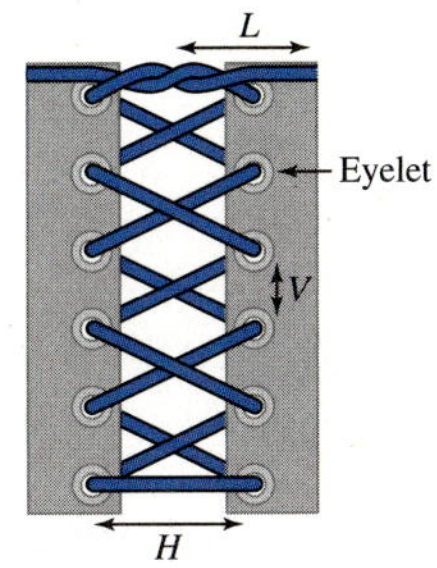

80. HOLIDAY DECORATIONS Find the length s of each string of colored lights used to decorate an evergreen tree in the manner shown if
$s = \sqrt{r^2 + h^2}$.

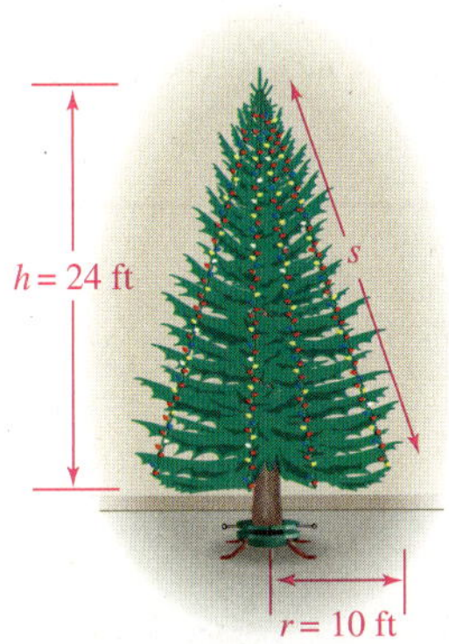

WRITING

81. When asked to find $\sqrt{16}$, a student answered 8. Explain his misunderstanding of the concept of square root.

82. Explain the difference between the *square* and the *square root* of a number. Give an example.

83. How would you check whether $\sqrt{389} = 17$?

84. Explain why $\sqrt{-4}$ is not equal to -2.

85. Is there a difference between $-\sqrt{25}$ and $\sqrt{-25}$? Explain.

86. A calculator was used to find $\sqrt{-16}$. Explain the meaning of the message on the display.

SECTION 1.11
Applications Introduction: The Pythagorean Theorem

In Section 1.11, we will study an important mathematical statement about right triangles called the **Pythagorean theorem**. (A **theorem** is a mathematical statement that can be proven.) This theorem expresses the relationship between the three sides of a **right triangle**. A right triangle is a triangle that has a **right angle** (an angle with measure 90º).

1. Which of the following appear to be right angles?

a. 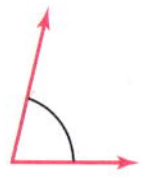

b. 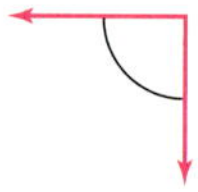

c. 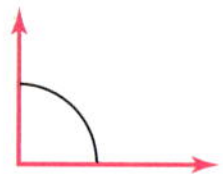

d. 

e. 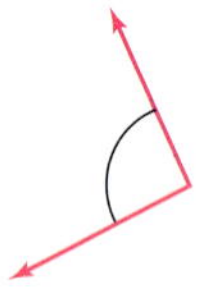

f. 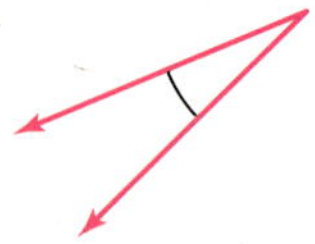

2. Each right triangle shown below has large red squares drawn on each of its three sides. Count the number of small squares inside each of the large red squares. Fill in the blanks to discover a mathematical relationship between your answers.

a. ____ + ____ = ____

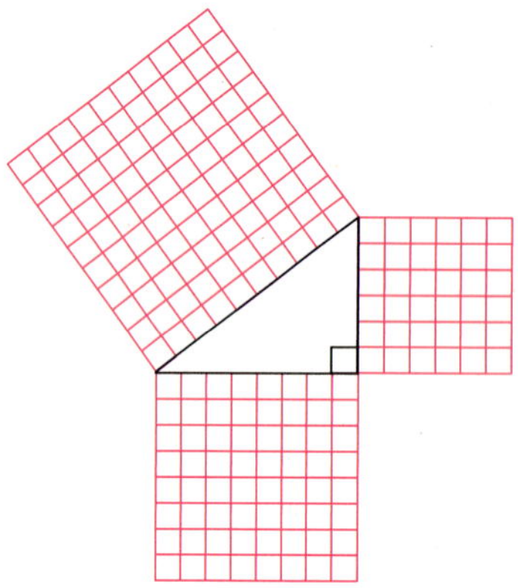

b. ____ + ____ = ____

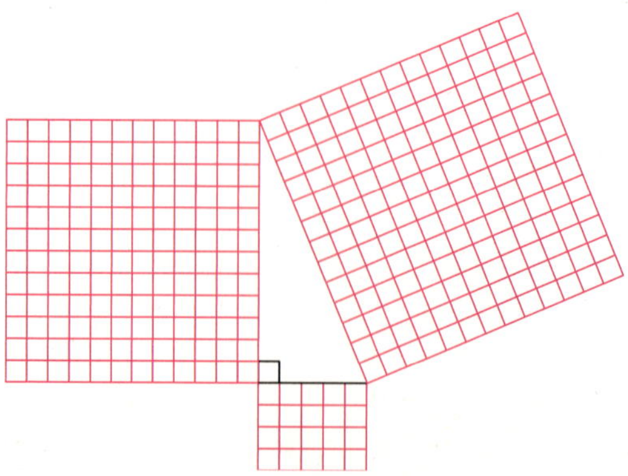

SECTION 1.11

The Pythagorean Theorem

Objectives

1. Classify angles and triangles.
2. Use the Pythagorean theorem to find the exact length of a side of a right triangle.
3. Use the converse of the Pythagorean theorem.

ARE YOU READY?

The following problems review some basic skills that are needed when working with right triangles.

1. Evaluate: $9^2 + 3^2$
2. How many sides does a triangle have?
3. Find each square root without using a calculator:
   a. $\sqrt{25}$ b. $\sqrt{225}$
4. Find each square root using a calculator:
   a. $\sqrt{1,369}$
   b. $\sqrt{222,784}$

Geometry is a branch of mathematics that studies the properties of two- and three-dimensional figures such as angles, triangles, circles, cylinders, and spheres. More than 5,000 years ago, Egyptian surveyors used geometry to measure areas of land in the flooded plains of the Nile River after heavy spring rains. Even today, engineers marvel at the Egyptians' use of geometry in the design and construction of the pyramids. History records many other practical applications of geometry made by Babylonians, Chinese, Indians, and Romans.

The Language of Algebra The word *geometry* comes from the Greek words *geo* (meaning earth) and *metron* (meaning measure).

1 Classify angles and triangles.

Geometry is based on three undefined words: point, line and plane. Other geometric figures can be created by using parts or combinations of points, lines, and planes. One such figure is called an angle. An angle is formed by two rays with a common endpoint. One unit of measurement of an angle is the degree. The symbol for degree is a small raised circle °. We classify angles according to their measure.

Classifying Angles

Acute angles: Angles whose measures are greater than 0° but less than 90°.

Right angles: Angles whose measures are 90°.

Obtuse angles: Angles whose measures are greater than 90° but less than 180°.

Straight angles: Angles whose measures are 180°.

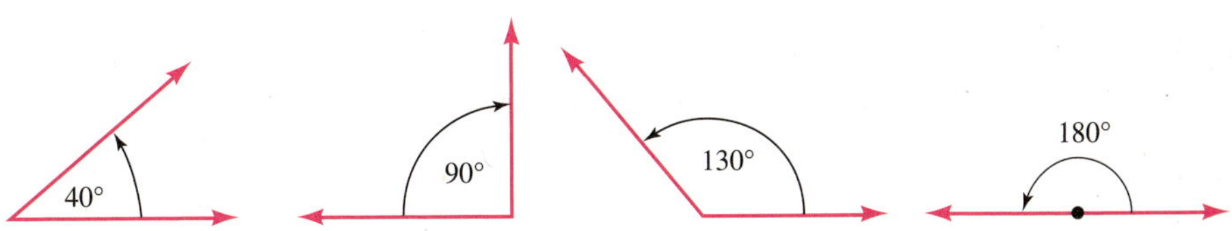

Acute angle **Right angle** **Obtuse angle** **Straight angle**

The Language of Algebra A ⌉ symbol is often used to label a right angle. For example, in the figure on the right, the ⌉ symbol indicates that the measure of the angle is 90°.

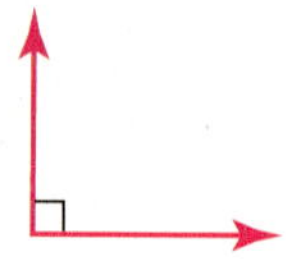

Triangles may also be classified by their angles, as shown below.

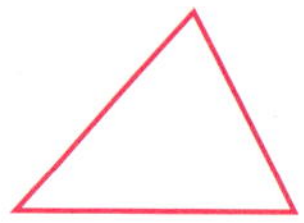

Acute triangle
(has three acute angles)

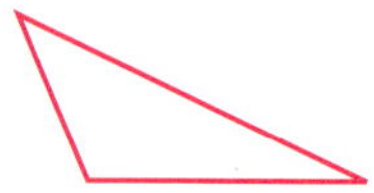

Obtuse triangle
(has an obtuse angle)

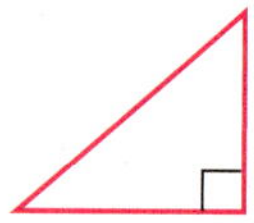

Right triangle
(has one right angle)

Pythagoras

2 Use the Pythagorean theorem to find the exact length of a side of a right triangle.

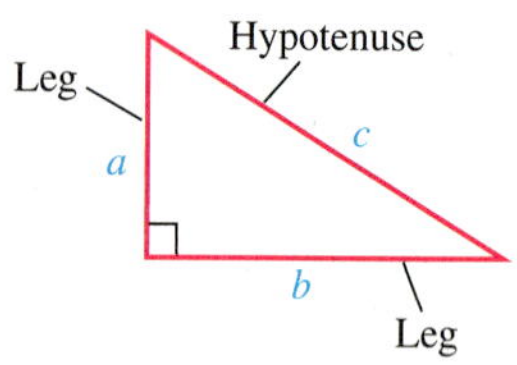

A **theorem** is a mathematical statement that can be proven. In this section, we will discuss one of the most widely used theorems of geometry—the Pythagorean theorem. It is named after Pythagoras, a Greek mathematician who lived about 2,500 years ago. He is thought to have been the first to develop a proof of it. The Pythagorean theorem expresses the relationship between the lengths of the sides of any right triangle. A right triangle is a triangle that has a right angle (an angle with measure 90°). In a right triangle, the longest side is called the **hypotenuse.** It is the side opposite the right angle. The other two sides are called **legs.** It is common practice to let the variable c represent the length of the hypotenuse and the variables a and b represent the lengths of the legs, as shown above.

If we know the lengths of any two sides of a right triangle, we can find the length of the third side using the **Pythagorean theorem.**

Pythagorean Theorem

If a and b are the lengths of two legs of a right triangle and c is the length of the hypotenuse, then

$$a^2 + b^2 = c^2$$

In words, the Pythagorean theorem is expressed as follows:

In a right triangle, the sum of the squares of the lengths of the two legs is equal to the square of the length of the hypotenuse.

Caution! When using the **Pythagorean equation** $a^2 + b^2 = c^2$, we can let a represent the length of either leg of the right triangle. We then let b represent the length of the other leg. The variable c must always represent the length of the hypotenuse.

EXAMPLE 1 Find the length of the hypotenuse of the right triangle shown here.

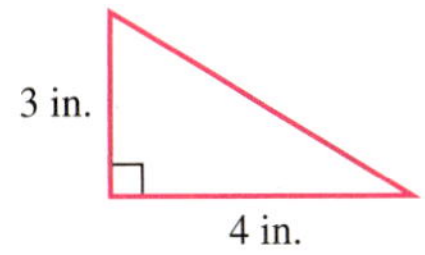

Strategy We will use the Pythagorean theorem to find the length of the hypotenuse.

WHY If we know the lengths of any two sides of a right triangle, we can find the length of the third side using the Pythagorean theorem.

Solution We will let $a = 3$ and $b = 4$, and substitute into the Pythagorean equation to find c.

$a^2 + b^2 = c^2$ This is the Pythagorean equation.

$3^2 + 4^2 = c^2$ Substitute 3 for *a* and 4 for *b*.

$9 + 16 = c^2$ Evaluate each exponential expression.

$25 = c^2$ Do the addition.

$c^2 = 25$ Reverse the sides of the equation so that c^2 is on the left.

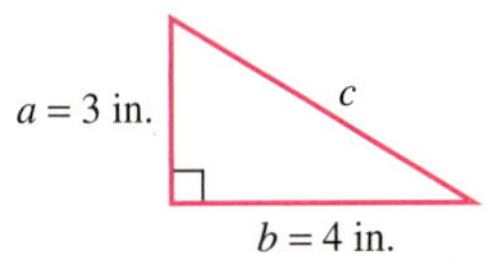

To find *c*, we must find a number that, when squared, is 25. There are two such numbers, one positive and one negative; they are the square roots of 25. Since *c* represents the length of a side of a triangle, *c* cannot be negative. For this reason, we need only find the positive square root of 25 to get *c*.

$c = \sqrt{25}$ The symbol $\sqrt{}$ is used to indicate the positive square root of a number.

$c = 5$ $\sqrt{25} = 5$ because $5^2 = 25$.

The length of the hypotenuse is 5 in.

Self Check 1

Find the length of the hypotenuse of the right triangle shown below.

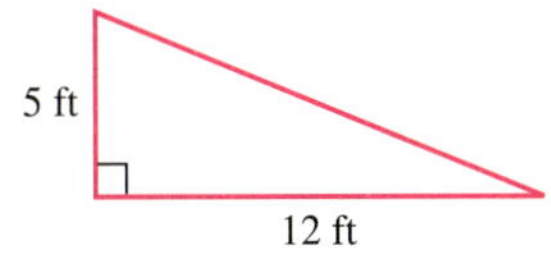

Now Try **Problem 17**

Success Tip The Pythagorean theorem is used to find the lengths of sides of right triangles. A calculator with a square root key $\boxed{\sqrt{}}$ is often helpful in the final step of the solution process when we must find the positive square root of a number.

EXAMPLE 2 ***Firefighting*** To fight a forest fire, the forestry department plans to clear a rectangular fire break around the fire, as shown in the following figure. Crews are equipped with mobile communications that have a 3,000-yard range. Can crews at points *A* and *B* remain in radio contact?

Strategy We will use the Pythagorean theorem to find the distance between points *A* and *B*.

WHY If the distance is less than 3,000 yards, the crews can communicate by radio. If it is greater than 3,000 yards, they cannot.

Solution The line segments connecting points *A*, *B*, and *C* form a right triangle. To find the distance *c* from point *A* to point *B*, we can use the Pythagorean equation, substituting 2,400 for *a* and 1,000 for *b* and solving for *c*.

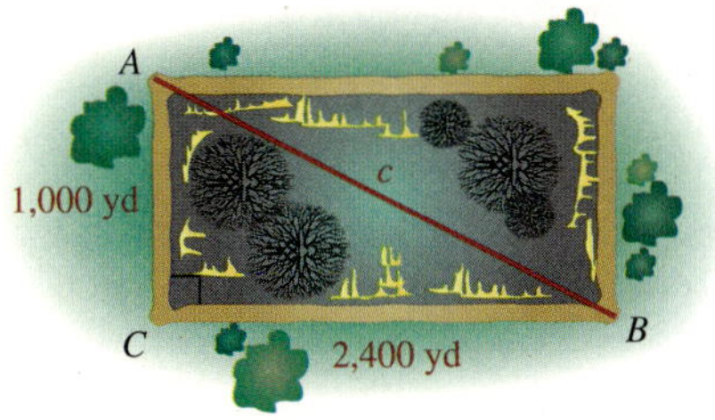

$a^2 + b^2 = c^2$ This is the Pythagorean equation.

$2{,}400^2 + 1{,}000^2 = c^2$ Substitute 2,400 for *a* and 1,000 for *b*.

$5{,}760{,}000 + 1{,}000{,}000 = c^2$ Evaluate each exponential expression.

$6{,}760{,}000 = c^2$ Do the addition.

Self Check 2

In Example 2, can the crews communicate by radio if the distance from point *B* to point *C* remains the same but the distance from point *A* to point *C* increases to 2,520 yards?

Now Try **Problems 21 and 29**

$c^2 = 6{,}760{,}000$ Reverse the sides of the equation so that c^2 is on the left.

$c = \sqrt{6{,}760{,}000}$ If $c^2 = 6{,}760{,}000$, then c must be a square root of 6,760,000. Because c represents a length, it must be the positive square root of 6,760,000.

$c = 2{,}600$ Use a calculator to find the square root.

The two crews are 2,600 yards apart. Because this distance is less than the 3,000-yard range of the radios, they can communicate by radio.

3 Use the converse of the Pythagorean theorem.

If a mathematical statement is written in the form *if p . . . , then q . . .* , we call the statement *if q . . . , then p . . .* its **converse.** The converses of some statements are true, while the converses of other statements are false. It is interesting to note that the converse of the Pythagorean theorem is true.

Converse of the Pythagorean Theorem

If a triangle has three sides of lengths a, b, and c, such that $a^2 + b^2 = c^2$, then the triangle is a right triangle.

Self Check 3

Is the triangle below a right triangle?

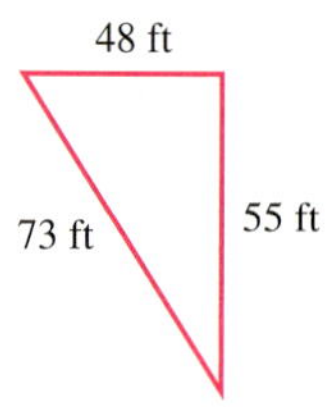

Now Try **Problem 25**

EXAMPLE 3

Is the triangle shown here a right triangle?

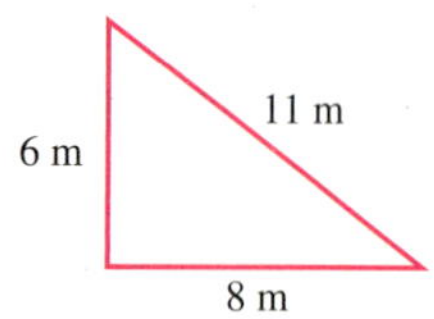

Strategy We will substitute the side lengths, 6, 8, and 11, into the Pythagorean equation $a^2 + b^2 = c^2$.

WHY By the converse of the Pythagorean theorem, the triangle is a right triangle if a true statement results. The triangle is not a right triangle if a false statement results.

Solution We must substitute the longest side length, 11, for c, because it is the possible hypotenuse. The lengths of 6 and 8 may be substituted for either a or b.

$a^2 + b^2 = c^2$ This is the Pythagorean equation.

$6^2 + 8^2 \stackrel{?}{=} 11^2$ Substitute 6 for a, 8 for b, and 11 for c.

$36 + 64 \stackrel{?}{=} 121$ Evaluate each exponential expression.

$100 = 121$ This is a false statement.

$$\begin{array}{r} \overset{1}{3}6 \\ +64 \\ \hline 100 \end{array}$$

Since $100 \neq 121$, the triangle is not a right triangle.

ANSWERS TO SELF CHECKS

1. 13 ft **2.** no **3.** yes

SECTION 1.11 STUDY SET

VOCABULARY

Fill in the blanks.

1. In a right triangle, the side opposite the 90° angle is called the __________. The other two sides are called _____.

2. The Pythagorean theorem is named after the Greek mathematician, __________, who is thought to have been the first to prove it.

3. The __________ theorem states that in any right triangle, the square of the length of the hypotenuse is equal to the sum of the squares of the lengths of the two legs.

4. $a^2 + b^2 = c^2$ is called the Pythagorean ________.

CONCEPTS

Fill in the blanks.

5. Classify each angle as an acute, right, obtuse, or straight angle.

a.

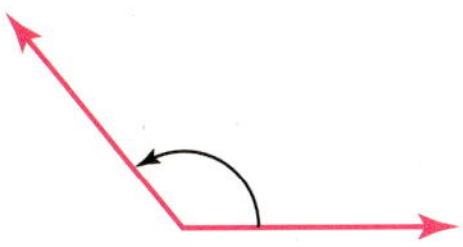

b.

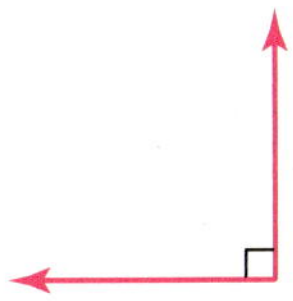

c.

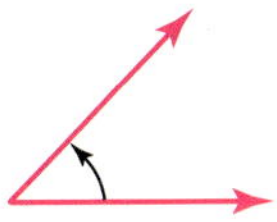

d.

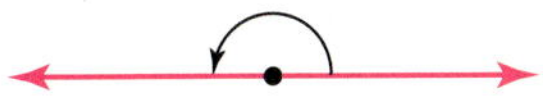

6. Draw an example of each type of angle.

a. an acute angle **b.** an obtuse angle

c. a right angle **d.** a straight angle

7. Classify each triangle as an acute, an obtuse, or a right triangle.

a.

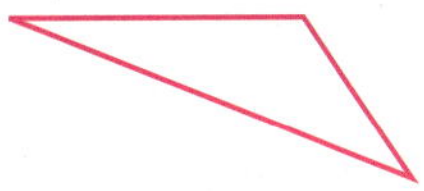

b.

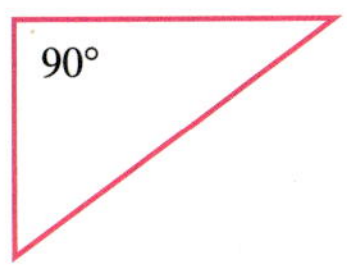

c.

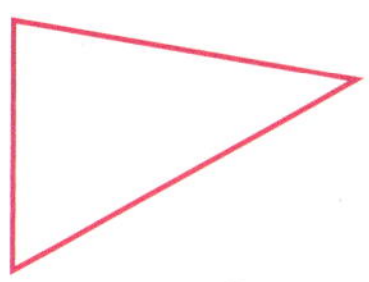

d.

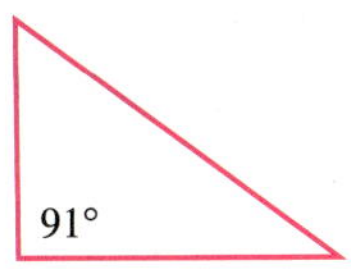

8. Draw an example of each type of triangle.

a. obtuse **b.** right

c. acute

9. If a and b are the lengths of two legs of a right triangle and c is the length of the hypotenuse, then ____ + ____ = ____.

10. The two solutions of $c^2 = 36$ are $c =$ ____ or $c =$ ____. If c represents the length of the hypotenuse of a right triangle, then we must discard the solution ____.

11. The converse of the Pythagorean theorem: If a triangle has three sides of lengths a, b, and c, such that $a^2 + b^2 = c^2$, then the triangle is a _____ triangle.

12. Refer to the triangle below.

a. What is the length of the hypotenuse?

b. What are the lengths of the two legs?

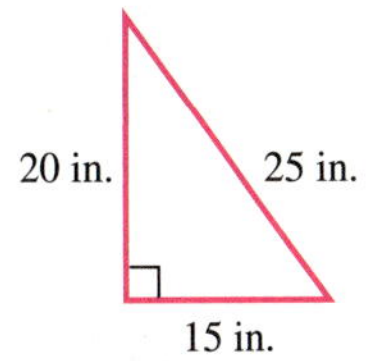

NOTATION

Complete the steps to solve the equation, where $a > 0$ and $c > 0$.

13.
$$8^2 + 6^2 = c^2$$
$$\_\_\_ + 36 = c^2$$
$$\_\_\_ = c^2$$
$$\sqrt{\_\_\_} = c$$
$$10 = c$$

Complete the steps.

14. The legs of a right triangle measure 5 and 12 centimeters. Find the length of the hypotenuse.

$$a^2 + b^2 = c^2$$
$$\_\_\_^2 + 12^2 = c^2$$
$$25 + \_\_\_ = c^2$$
$$\_\_\_ = c^2$$
$$\sqrt{169} = \_\_\_$$
$$\_\_\_ = c$$

The length of the hypotenuse is ____ cm.

GUIDED PRACTICE

Find the length of the hypotenuse of the right triangle shown below if it has the given side lengths. See Examples 1 and 2.

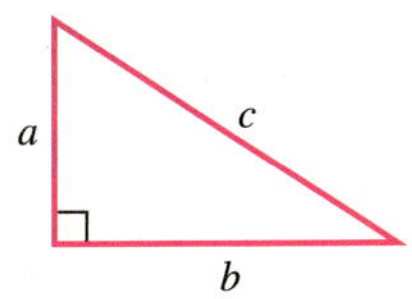

15. $a = 6$ ft and $b = 8$ ft

16. $a = 12$ mm and $b = 9$ mm

17. $a = 5$ m and $b = 12$ m

18. $a = 16$ in. and $b = 12$ in.

19. $a = 48$ mi and $b = 55$ mi

20. $a = 80$ ft and $b = 39$ ft

21. $a = 88$ cm and $b = 105$ cm

22. $a = 132$ mm and $b = 85$ mm

Is a triangle with the following side lengths a right triangle?
See Example 3.

23. 12, 14, 15

24. 15, 16, 22

25. 33, 56, 65

26. 20, 21, 29

APPLICATIONS

Pythagorean Theorem Problems

27. HIGH-ROPES ADVENTURES COURSES A builder of a high-ropes adventure course wants to secure a pole by attaching a support cable from the anchor stake 8 yards from its base to a point 6 yards up the pole. How long should the cable be?

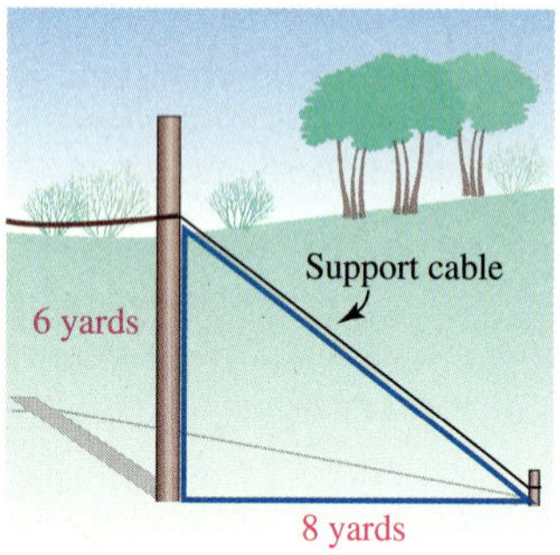

28. TIGHTROPE WALKERS A circus performer intends to walk up a taut cable shown in the illustration to a platform at the top of a pole. How long is the cable?

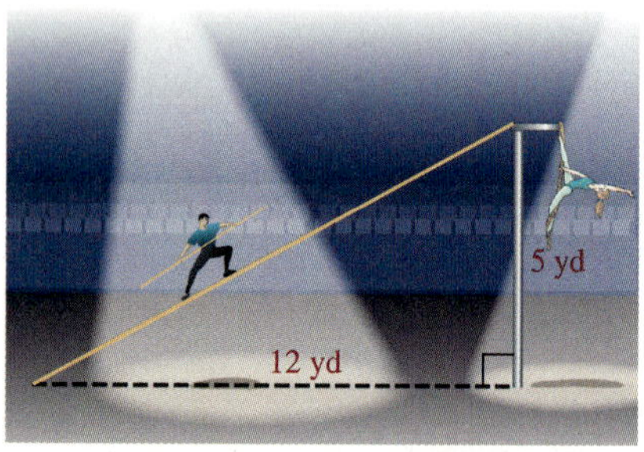

29. PICTURE FRAMES After gluing and nailing two pieces of picture frame molding together, a frame maker checks her work by making a diagonal measurement. If the sides of the frame form a right angle, what measurement should the frame maker read on the yardstick?

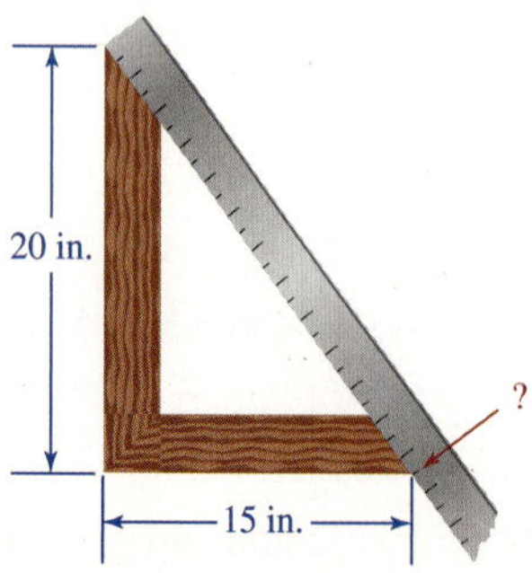

30. CARPENTRY The gable end of the roof shown is divided in half by a vertical brace, 8 feet in height. Find the length of the roof line.

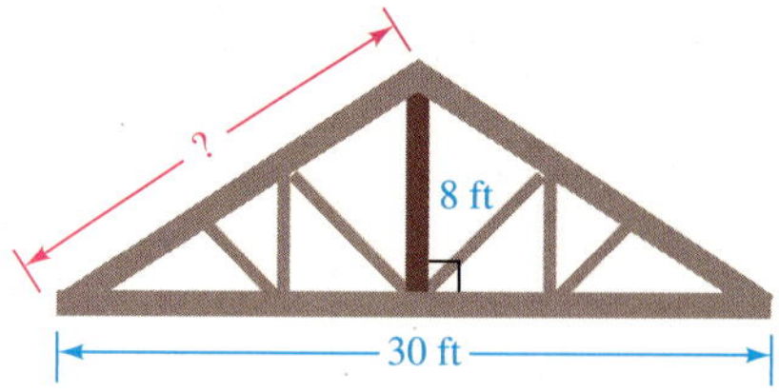

31. MOTO X Find the length of the landing ramp.

32. LCD FLAT PANEL HD SCREEN The size of an HD screen is the diagonal distance from the upper left to the lower right corner. What is the size of the screen shown in the illustration?

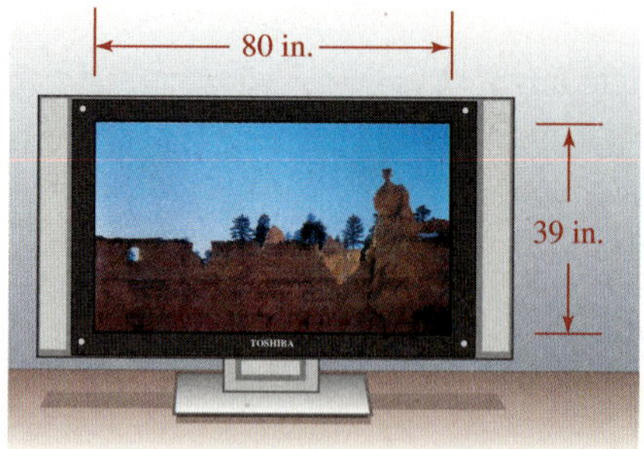

33. CARPENTRY A carpenter's square is made of flat steel or aluminum and is shaped like an L. It is used for framing, roofing, and stairway work.

a. How long is each side of the carpenter's square shown below? (The units are inches.)

b. What should be the diagonal measurement (shown in red) that will guarantee that the sides of the square form a 90° angle?

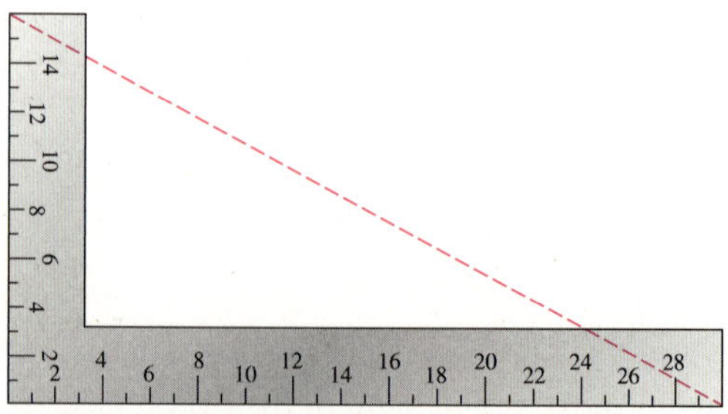

34. WIND DAMAGE A tree was blown over in a wind storm. Find the height of the tree when it was standing vertically upright.

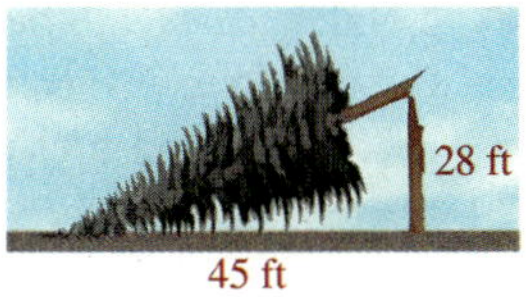

35. SHIPPING CRATES Find the length of the diagonal brace on the shipping crate in the illustration.

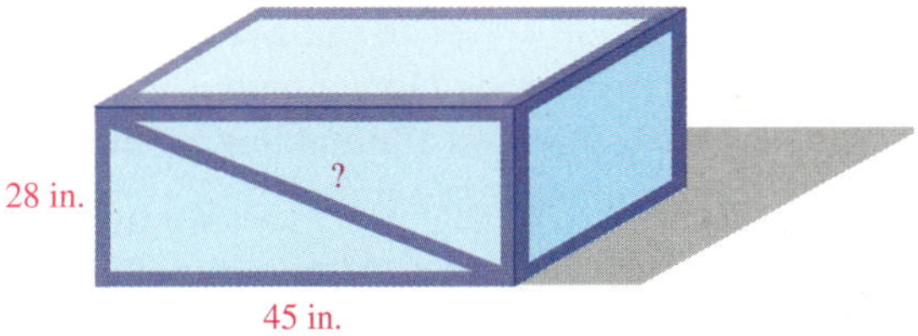

36. CARGO SPACE How wide a piece of plywood can be stored diagonally in the back of the van?

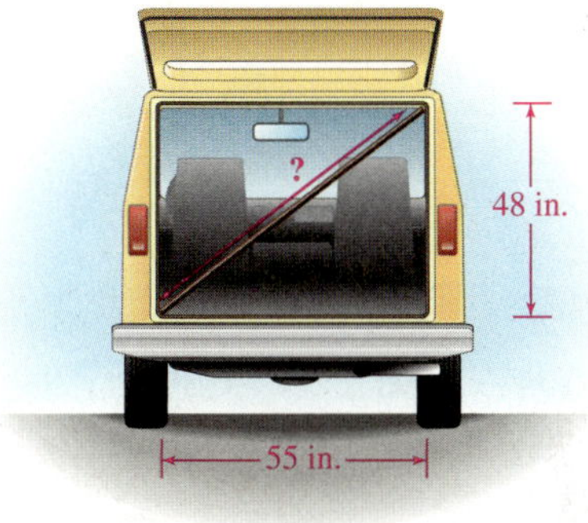

37. SHORTCUTS Instead of walking on the sidewalk, students take a diagonal shortcut across a rectangular vacant lot, as shown in the illustration. How much distance do they save?

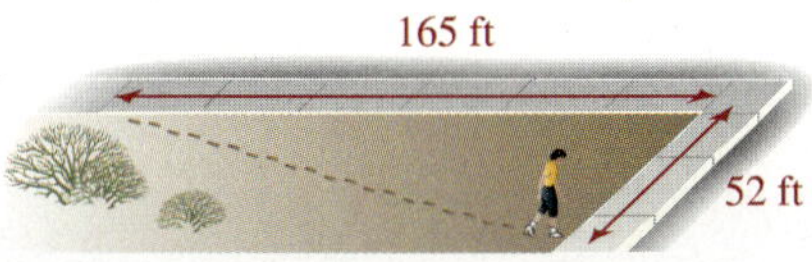

38. SAILING A sailboat cannot sail directly into the wind. In order to sail upwind, the boat must travel back and forth across the wind using a technique called *tacking*. In the illustration below, a sailboat follows the two-part tacking course shown in blue. For how many yards must the sailboat tack to eventually advance 88 yards into the wind?

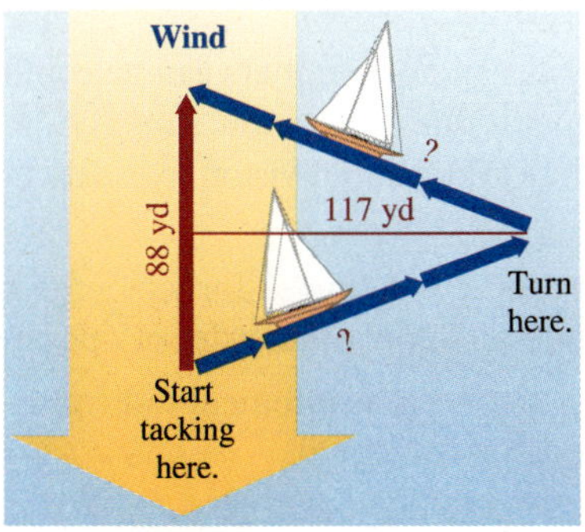

39. HOME FURNISHINGS A 36-inch by 48-inch opening in an oak wall unit is designed to hold a television screen. What is the largest size screen that will fit in the opening? (*Hint:* The size of a television screen is the diagonal distance from the upper left to lower right corner.)

40. COLLISIONS A driver is traveling on a street that is perpendicular to some railroad tracks. He is distracted and doesn't see a railroad crossing up ahead. Meanwhile, a slow-moving train is headed toward that same crossing. At the instant the car is 15 feet from the crossing, the train's collision detector alarm sounds and the readout indicates that the train is 112 feet away.

a. Draw a diagram of this situation and label the given distances.

b. At the instant the train's collision alarm sounds, how far apart are the train and car?

41. TREASURE ISLAND The directions for finding the location of buried treasure are given below:

From the tallest date palm tree on the beach, walk 28 paces due north. Then turn and walk due east 45 paces. At that location, dig down 6 feet. That's where you'll find the treasure.

a. Draw a diagram of this situation that shows the position of the date palm tree and where to dig on the beach. Label the given distances.

b. What is the distance between the location to dig and the date palm tree?

42. DESERT OFF-ROADING The driver of an off-road vehicle left base camp and traveled 7 miles due west. From that point, he drove due north for 24 miles. "As the crow flies," how far away from base camp was he at that stage of the trip?

WRITING

43. State the Pythagorean theorem in your own words.

44. When the lengths of the sides of the triangle shown below are substituted into the equation $a^2 + b^2 = c^2$, the result is a false statement. Explain why.

$$a^2 + b^2 = c^2$$
$$2^2 + 4^2 = 5^2$$
$$4 + 16 = 25$$
$$20 = 25$$

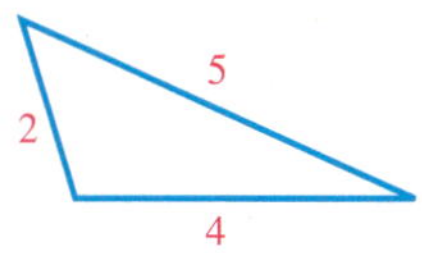

45. In the figure on the next page, equal-sized squares have been drawn on the sides of right triangle $\triangle ABC$. Explain how this figure demonstrates that $3^2 + 4^2 = 5^2$.

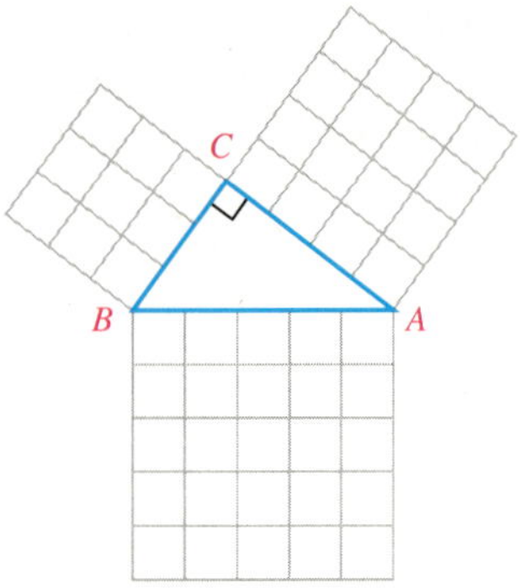

46. CARPENTRY A carpenter uses a tape measure to see whether the wall he just put up is perfectly square with the floor. Explain what mathematical concept he is applying. If the wall is positioned correctly, what should the measurement on the tape read?

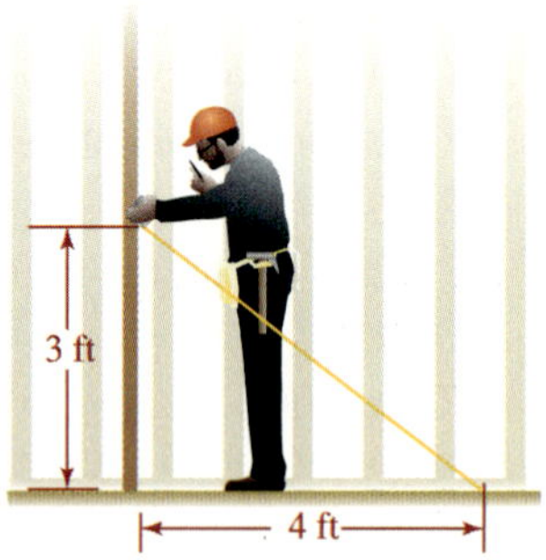

47. What error is apparent in the following illustration?

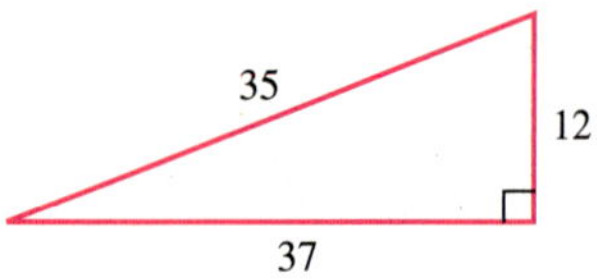

48. When naming the legs of a right triangle, explain why it doesn't matter which leg you label a and which leg you label b.

49. THE WIZARD OF OZ In the 1939 classic movie, the Scarecrow was in search of a brain. Once he received an honorary degree in "Thinkology" from the Wizard, he tried to impress his friends by stating: "*The sum of the square roots of any two sides of an isosceles triangle is equal to the square root of the remaining side.*" (You can watch this scene on YouTube.) What well-known mathematical fact was the Scarecrow attempting to recite? Explain the errors that he made. (Over 50 years later, in an episode of *The Simpsons,* Homer quotes the Scarecrow word-for-word after finding a pair of eyeglasses in a public restroom.)

©Pictorial Press Ltd/Alamy

50. **a.** What is an acute angle?

b. What is an obtuse angle?

c. What is a right angle?

MODULE 1 TEST

1. Fill in the blanks.

a. $\{\ldots, -5, -4, -3, -2, -1, 0, 1, 2, 3, 4, 5, \ldots\}$ is called the set of ________.

b. The symbols $>$ and $<$ are called ________ symbols.

c. The ________ ________ of a number is the distance between the number and 0 on the number line.

d. ________ are letters (or symbols) that stand for numbers.

e. In the expression $(-3)^5$, the ______ is -3 and 5 is the ________.

f. Variables and/or numbers can be combined with the operations of addition, subtraction, multiplication, and division to create algebraic ________.

g. To evaluate $y^2 + 9y - 3$ for $y = -5$, we ________ -5 for y and apply the order of operations rule.

h. An ________ is a statement indicating that two expressions are equal.

i. A ______ is an equation that states a mathematical relationship between two or more variables.

j. The distance around a plane geometric figure is called its ________, and the amount of surface that it encloses is called its ____.

k. An _______ triangle is a triangle with two sides of the same length.

l. The symbol $\sqrt{\ }$ is called a ______ symbol. It represents the ______ or principal square root of a number. The symbol $-\sqrt{\ }$ is used to represent the _______ square root of a number.

m. A ____ triangle is a triangle that has a 90° angle.

n. The longest side of a right triangle is the ________. The remaining two sides are the ____ of the triangle.

o. The _________ theorem is a formula that relates the lengths of the three sides of a right triangle.

2. Insert one of the symbols > or < in the blank to make the statement true.

a. -8 ____ -9 **b.** -213 ____ 123 **c.** -5 ____ 0

3. Tell whether each statement is true or false.

a. $19 \geq 19$ **b.** $-(-8) = 8$

c. $-|-2| > |6|$ **d.** $-7 + 0 = 0$

e. $-5(0) = 0$

4. SCHOOL ENROLLMENT According to the projections in the table, which high school will face the greatest shortage of classroom seats in the year 2020?

High Schools with Shortage of Classroom Seats by 2020

Lyons	−669
Tolbert	−1,630
Poly	−2,488
Cleveland	−350
Samuels	−586
South	−2,379
Van Owen	−1,690
Twin Park	−462
Heywood	−1,004
Hampton	−774

5. Graph the following numbers on a number line: -3, the additive inverse of -4, the opposite of 1, and 3

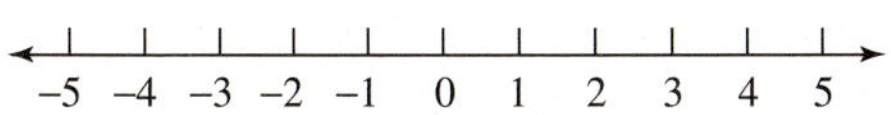

6. Let a represent a number greater than zero and b represent a number less than 0. Determine whether each statement is *sometimes true*, *always true*, or *never true*. Explain your reasoning.

a. $a > b$ **b.** $b < -b$

c. $|a| < |b|$ **d.** $|b| < 0$

7. Translate the following phrase to mathematical symbols: *Negative twenty-one minus negative seventy-three.*

8. Illustrate each addition of integers below using a model. Ask your instructor which model (heaps/holes, colored tiles, number line, or money) you should use. Then write a number sentence of the form ____ + ____ = ____ to express the result in mathematical symbols.

a. $6 + (-2)$

b. $-7 + 3$

c. $1 + (-6)$

d. $-4 + (-5)$

9. a. Illustrate the following process using a number line model. Begin at 0 and move 4 units to the right. Then move 4 units to the left. On what integer does the process end?

b. Illustrate the addition problem described in part a using a heaps/holes, colored tiles, or money model. (Ask your instructor which one you should use.)

c. Write a number sentence of the form ____ + ____ = ____ to express your result to part b in mathematical symbols.

10. Add.

a. $-6 + 3$ **b.** $-72 + (-73)$

c. $8 + (-6) + (-9) + 5 + 1$

d. $(-31 + 12) + [3 + (-16)]$

e. $-24 + (-3) + 24 + (-5) + 5$

11. Subtract.

a. $-7 - 6$ **b.** $-7 - (-6)$

c. $82 - (-109)$ **d.** $0 - 15$

e. $-60 - 50 - 40$

12. Multiply.

a. $-10 \cdot 7$ **b.** $-4(-73)$

c. $-4(2)(-6)$ **d.** $-9(-3)(-1)(-2)$

e. $-20{,}000(1{,}300)$

13. Write the related multiplication statement for $\frac{-20}{-4} = 5$.

14. Divide and check the result.

a. $\frac{-32}{4}$ **b.** $24 \div (-3)$

c. $-54 \div (-6)$ **d.** $\frac{408}{-12}$

e. $-560{,}000 \div 7{,}000$

15. a. What is 15 more than -27?

b. Subtract -19 from -1.

c. Divide -28 by -7.

d. Find the product of 10 and the opposite of 8.

Find two integers that meet the given requirement. Express your answer in a number sentence using mathematical symbols. For example, $-1 + 3 = -2$.

16. a. The sum of two integers is -11, and only one of them is negative.

b. The sum of a positive integer and a negative integer is a positive integer.

c. The difference of two negative integers is -9.

d. The quotient of two integers is -48.

17. Find two integers whose sum is -1 and whose product is -30.

18 a. What property is shown: $-3 + 5 = 5 + (-3)$

b. What property is shown: $-4(-10) = -10(-4)$

c. Fill in the blank: Subtracting is the same as _______ the opposite.

19. Divide, if possible.

a. $\frac{-21}{0}$ **b.** $\frac{-5}{1}$

c. $\frac{0}{-6}$ **d.** $\frac{-18}{-18}$

20. a. What property of addition can be used to make the evaluation of $(-453 + 175) + 25$ easier?

b. Evaluate the expression using your answer to part a.

21. Evaluate each expression:

a. $(-4)^2$ **b.** -4^2

Evaluate each expression.

22. $4 - (-3)^2 - (-6)$ **23.** $-18 \div 2 \cdot 3$

24. $-3 + \left(\frac{-16}{4}\right) - 3^3$

25. $94 - 3[-7 + (5 - 8)^2]$

26. $\frac{4(-6) - 4^2 + (-2)}{-3 - 4 \cdot 1^5}$

27. $6(-2 \cdot 6 + 5 \cdot 4)$

28. $21 - 9|-3 - 4 + 2|$

29. $-\left[2 - \left(4^3 + \frac{20}{-5}\right)\right]$

30. CHEMISTRY In a lab, the temperature of a fluid was reduced 6°F per hour for 12 hours. What signed number represents the change in temperature?

31. GAMBLING On the first hand of draw poker, a player won the chips shown on the left. On the second hand, he lost the chips shown on the right. Determine his net gain or loss for the first two hands. The dollar value of each colored poker chip is shown.

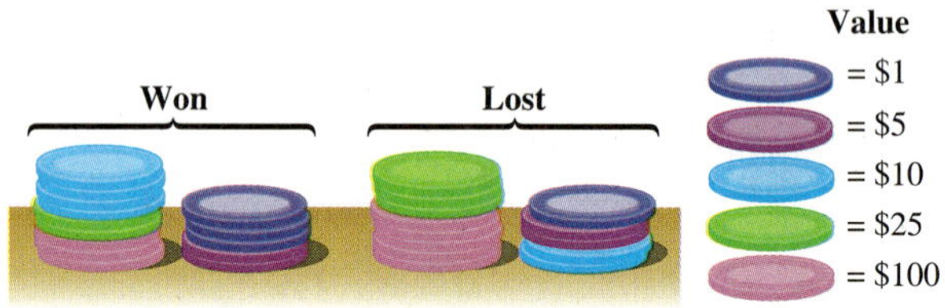

32. GEOGRAPHY The lowest point on the African continent is the Qattarah Depression in the Sahara Desert, 436 feet below sea level. The lowest point on the North American continent is Death Valley, California, 282 feet below sea level. Find the difference in these elevations.

33. TRAMS A tram line makes a 5,250-foot descent from a mountaintop to the base of the mountain in 15 equal stages. How much does it descend in each stage?

34. CARD GAMES After the first round of a card game, Tommy had a score of 8. When he lost the second round, he had to deduct the value of the cards left in his hand from his first-round score. (See the illustration.) What was his score after two rounds of the game? For scoring, face cards (Kings, Queens, and Jacks) are counted as 10 points and aces as 1 point.

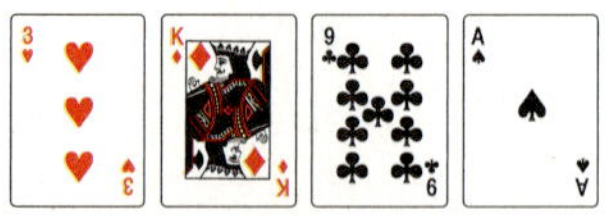

35. BANK TAKEOVERS Before three investors can take over a failing bank, they must repay the losses that the bank had over the past three quarters. If the investors plan equal ownership, how much of the bank's total losses is each investor responsible for?

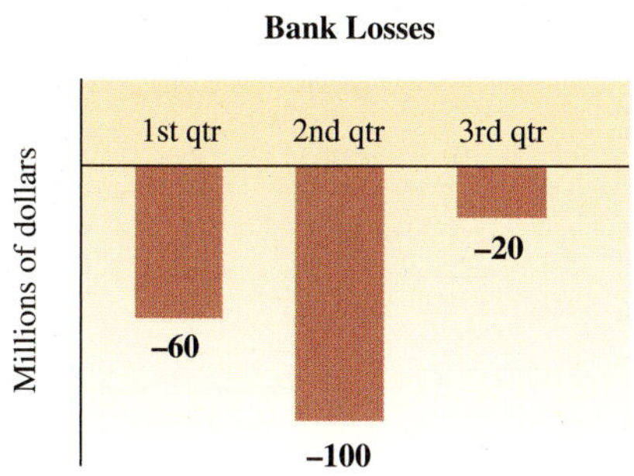

36. Find the values for points A, B, and C on the number line below. Then write a number sentence that calculates each change.

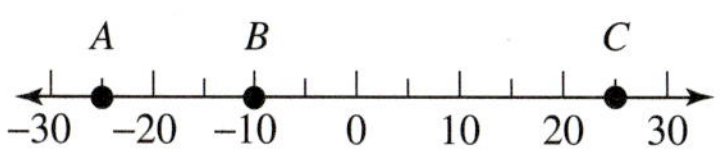

a. From C to A
b. From B to C

37. PLANETS Mercury orbits closer to the sun than any other planet. Temperatures on Mercury can get as high as 810°F and as low as −290°F. What is the temperature range?

38. OVERDRAFT PROTECTION A student forgot that she had only \$30 in her bank account and wrote a check for \$55 and used her debit card to buy \$75 worth of groceries. On each of the two transactions, the bank charged her a \$20 overdraft protection fee. Find the new account balance.

39. ***Classify each item as an algebraic expression or an equation.***

a. $m + 18 = 23$ **b.** $m + 18$
c. $30x$ **d.** $30x = 600$

40. What arithmetic operations does the expression $\frac{12 + 9t}{25}$ contain? What variable does it contain?

41. SALARIES A wife's monthly salary is \$1,000 less than twice her husband's monthly salary.

a. If her husband's monthly salary is h dollars, write an algebraic expression that represents the wife's monthly salary (in dollars).

b. Suppose the husband's monthly salary is \$2,350. Find the wife's monthly salary.

42. REFRESHMENTS Write an algebraic expression that represents the number of cups of coffee are left in the coffeemaker shown if c cups have already been poured from it.

43. Let a variable represent the length of one of the fish shown. Then write an expression that represents the length (in inches) of the other fish. Give two possible sets of answers.

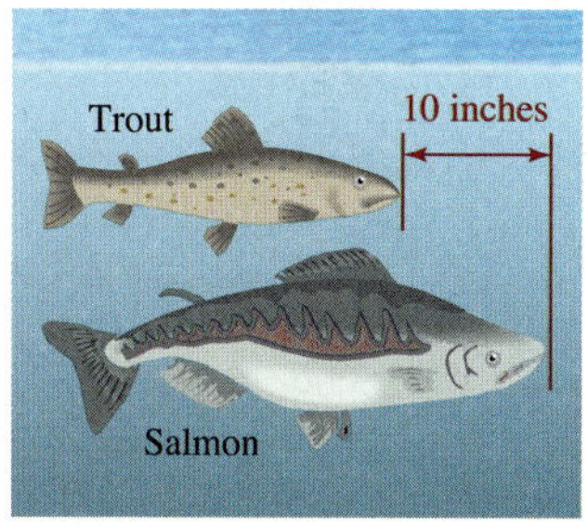

44. Translate each phrase to mathematical symbols.

a. 2 less than r
b. The product of 3, x, and y
c. x increased by 100
d. The absolute value of the quotient of x and -9

45. Write an algebraic expression that represents the number of years in d decades.

46. Evaluate each expression.

a. $x - 16$ for $x = 4$
b. $2t^2 - 3(t - s)$ for $t = -2$ and $s = 4$
c. $-a^2 + 10$ for $a = -3$
d. $\left|\frac{-10d + f^3}{-f}\right|$ for $d = -1$ and $f = -5$

47. DISTANCE TRAVELED Find the distance traveled by a motorist who departed from home at 9:00 A.M. and arrived at his destination at noon, traveling at a rate of 55 miles per hour.

48. PROFITS A craft show promoter had revenues and costs as shown. Find the profit.

Revenues	Costs
Ticket sales: \$40,000	Supplies: \$13,000
Booth rental: \$15,000	Facility rental fee: \$5,000

49. FALLING OBJECTS If a tennis ball was dropped from the top of a 200-foot-tall building, would it hit the ground after falling for 3 seconds? If not, how far short of the ground would it be?

50. VEHICLE WEIGHTS A Hummer H2 weighs 340 pounds less than twice a Honda Element.

a. Let x represent the weight of one of the vehicles. Write an expression for the weight of the other vehicle.

b. If the weight of the Element is 3,370 pounds, what is the weight of the Hummer?

51. METER READINGS Every hour between 8 A.M. and 5 P.M., a technician noted the value registered by a meter in a power plant and recorded that number on a line graph. Find the mean (average) meter value reading for this period.

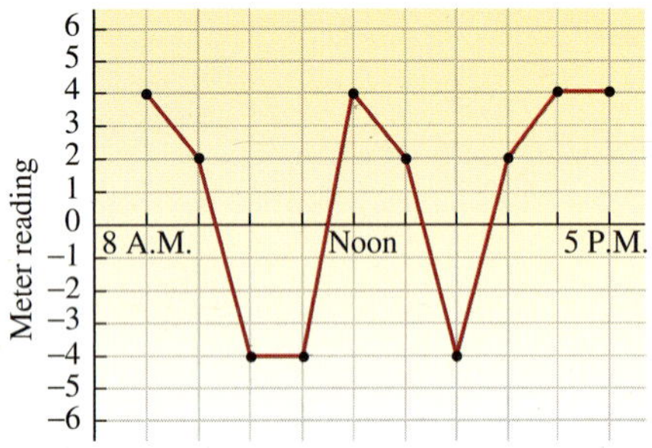

52. Determine which concept (perimeter or area) should be used to find each of the following. Then determine which unit of measurement, ft or ft^2, would be appropriate.

a. The amount of ground covered by a sleeping bag lying on the floor

b. The distance around a dance floor

53. Find the perimeter of a rectangle with length 45 in. and width 29 in.

54. Find the area of a square with sides of 60 ft long.

55. Find the area of the rectangle shown below in square inches.

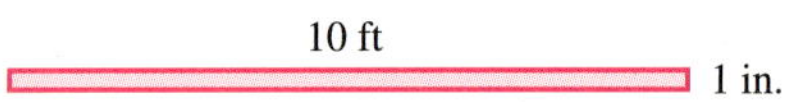

56. HOME DECORATING A rectangular living room measures 24 feet by 15 feet. At $28 per square yard (labor and materials), how much will it cost to carpet the room? Assume no waste.

57. Find the area of the shaded part of the figure shown below.

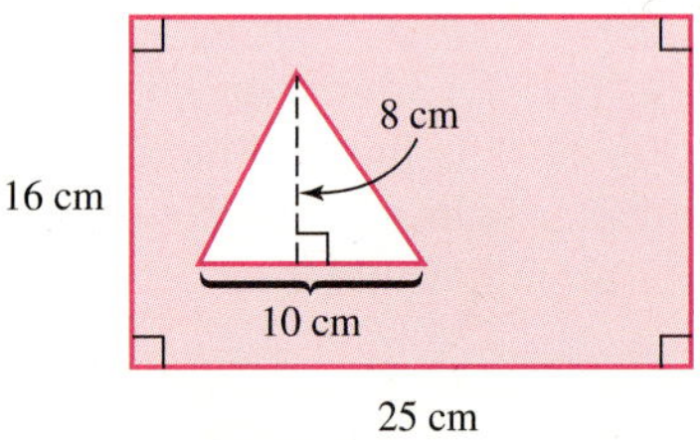

58. An isosceles triangle has sides of length 10 yds and 12 yds. What are the two possible perimeters that the triangle could have?

59. **a.** Sketch and label two rectangles each having an area of 90 ft^2.

b. Do the rectangles that you drew in part a have the same perimeter?

60. Find the area of a triangle with base 17 centimeters and height 28 centimeters.

61. Find the area of the figure.

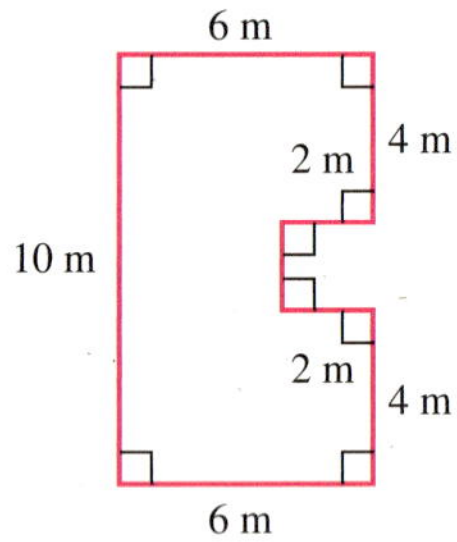

62. Evaluate each expression without using a calculator.

a. $-2\sqrt{25} + 3\sqrt{49}$

b. $8 - 2(2^4 - 60 + 6\sqrt{81})$

63. FALLING OBJECTS The time t (in seconds) that it takes for an object to fall a distance of s feet is given by the formula $t = \frac{\sqrt{s}}{4}$. How long would it take a stone dropped over the edge of a canyon that is 1,296 feet deep to hit bottom?

64. If the legs of a right triangle measure 10 meters and 24 meters, how long is the hypotenuse?

65. TECHNOLOGY What is the diagonal measurement of the television screen shown below?

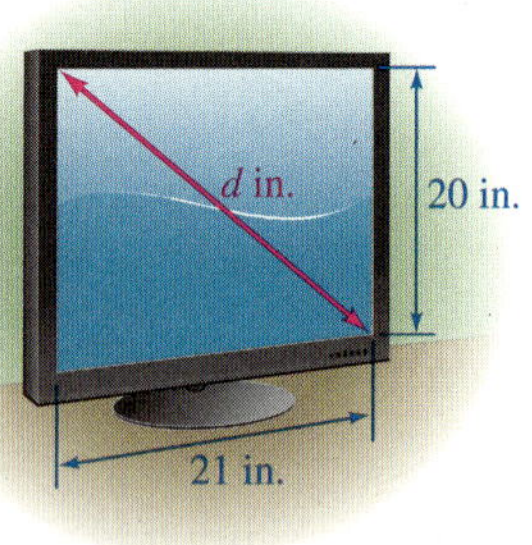

66. THEATER SEATING The seat at the top of the incline is 5 feet higher than the seat at the bottom. How long is the incline?

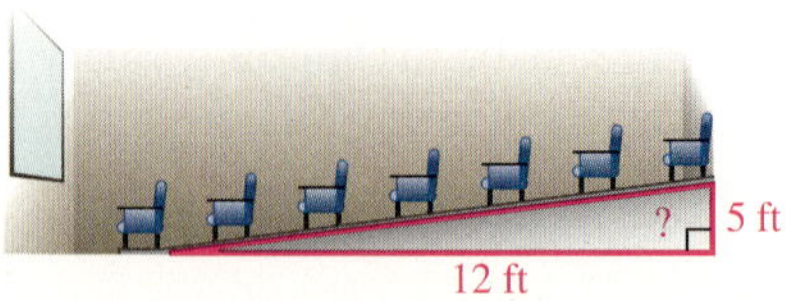

67. TIME CAPSULES The directions for finding the location of a time capsule buried 100 years ago in a field are given below.

From Founder's Rock, go 85 feet east. Then go north 132 feet. At that location, the capsule is 5 feet below the ground.

a. Draw a diagram of this situation that shows the positions of Founder's Rock and where to dig in the park. Label the given distances.

b. What is the distance between the location to dig and Founder's Rock?

68. Determine whether a triangle with sides of length 16 feet, 63 feet, and 65 feet is a right triangle.

Module 1 Answers

Applications Introduction : Integers Section 1.1 (page 4)

1. **a.** negative **b.** positive **c.** positive **d.** negative **e.** negative **f.** positive **g.** negative **h.** positive

2. **a.** 1 **b.** – 5 **c.** – 8 **d.** 8 **e.** 0 **f.** 0

Are You Ready? Section 1.1 (Page 5)

1. {0, 1, 2, 3, 4, 5, . . .} **2.** 0, 3, 7, 9, 15, 26

3. 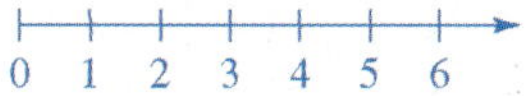

4. below

Study Set Section 1.1 (page 12)

1. Positive, negative

3. 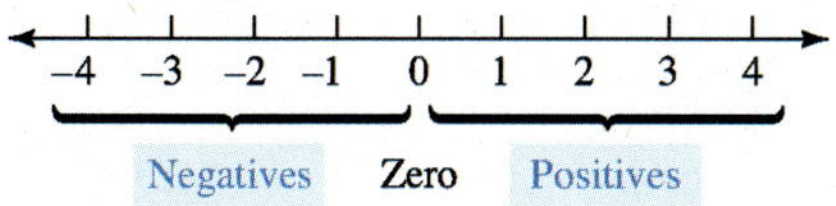

5. graph

7. absolute, value **9.** **a.** – 225 **b.** 15 lb **c.** – 10 sec **d.** – 3° **e.** 10° **f.** – $12,000 **g.** – 1 mi **h.** 15,257

11. 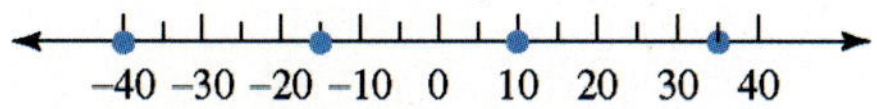

13. **a.** – 4 **b.** –2 **15.** **a.** –7 **b.** 8 **17.** **a.** 15 > – 12 **b.** –5 < – 4 **19.** 25, 25, –39, 39, 0, 0 **21.** **a.** – (–8) **b.** | –8 | **c.** 8 – 8 **d.** – | – 8 |

23. **a.** greater, equal **b.** less, equal **25.** braces

27. 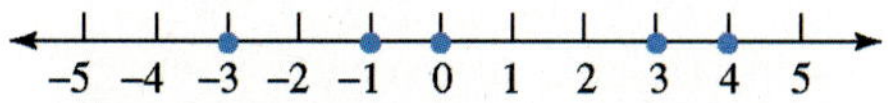

29. 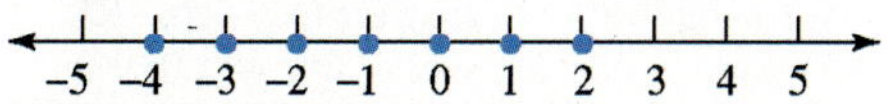

31. 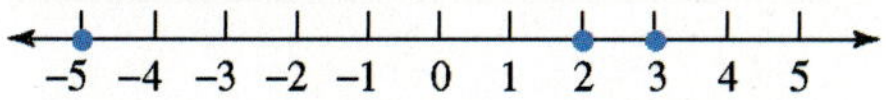

33. 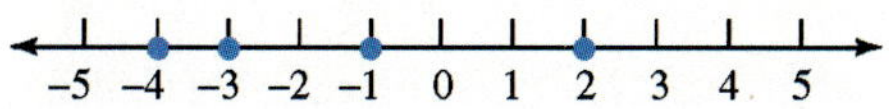

35. < **37.** < **39.** > **41.** > **43.** true **45.** true **47.** false **49.** false **51.** 9 **53.** 8 **55.** 14 **57.** 180 **59.** 11 **61.** 4 **63.** 102 **65.** 561 **67.** –20 **69.** –6 **71.** –253 **73.** 0 **75.** > **77.** < **79.** > **81.** < **83.** –52, –22, –12, 12, 52, 82 **85.** –3, –5, –7 **87.** **a.** < **b.** > **c.** >, < **d.** > **89.** 1,999 **91.** – 31 Lengths

93. 0, 20, 5, – 40, – 120 **95.** peaks: 2, 4, 0; valleys:–3, –5, –2 **97.** **a.** – 1 (1 below par) **b.** – 3 (3 below par) **c.** Most of the scores are below par.

99. **a.** 10° to 20° **b.** –20° to –10° **c.** 40° **d.** 10°

101. **a.** 200 yr **b.** A.D. **c.** B.C. **d.** the birth of Christ

103. 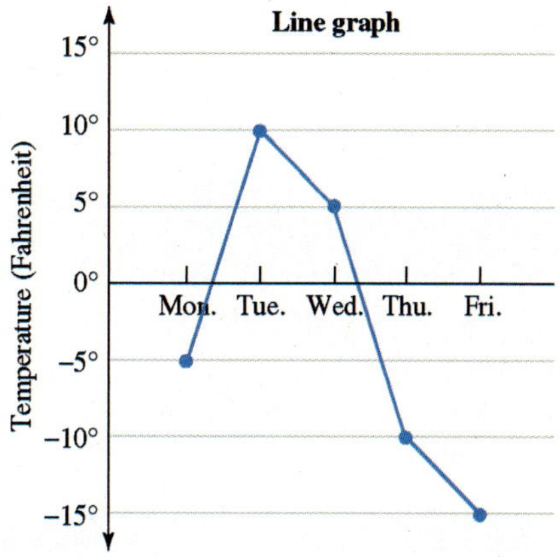

105. iii

Applications Introduction: Adding Integers Using Models Section 1.2 (page 19)

1. **a.** 4, (– 3), 1 **b.** – 2, 6, 4 **2.** **a.** 4 **b.** –3 **c.** 0 **d.** – 11 **e.** – 60 **f.** 500 **g.** – 2 **3.** **a.** 2 **b.** 2; Yes

4. **a.** Same result **b.** Not the same result **c.** Not the same result **d.** Same result **e.** Same result **f.** Not the same result

5. **a.** – 8, 3, – 5 **b.** 6, (– 2), 4 **c.** 3 **d.** 2 **6.** **a.** 5 **b.** – 2 **c.** – 5 **d.** 0 **e.** – 40 **f.** 200 **g.** 1 **7.** **a.** positive **b.** negative **c.** negative **d.** positive **8.** **a.** 2, 5 **b.** – 4, – 6 **c.** – 6, – 2 **d.** 9, 4 **9.** **a.** positive **b.** negative **c.** negative **d.** positive **10.** **a.** ahead by $20 **b.** behind by $20 **c.** ahead by $80 **d.** behind by $80 **e.** behind by $20 **f.** ahead by $80 **g.** behind by $80

Are You Ready? Section 1.2 (page 22)

1. 3, 5; –5 has the larger absolute value **2.** 11 **3.** 621 **4.** 9

Study Set Section 1.2 (page 31)

1. like, unlike **3.** identity **5.** Commutative **7.** **a.** 1 ft deep **b.** – 6 + 5 = – 1 **9.** **a.** 5 + (– 2) = 3 **b.** – 8 + 3 = – 5 **c.** 2 + (– 6) = –4 **d.** – 4 + (– 5) = – 9 **11.** **a.** 0 **c.** 3 + (– 3) = 0 **13.** subtract, larger **15.** **a.** yes **b.** yes **c.** no **d.** no **17.** **a.** 0 **b.** 0 **19.** **a.** 1, (– 5) **b.** – 80.5, 15 **c.** 20, 4 **d.** 3, 2.1 **21.** – 10, – 4 **23.** – 9 **25.** – 10 **27.** – 62 **29.** – 96 **31.** – 379 **33.** – 874 **35.** – 3 **37.** 1 **39.** – 22 **41.** 48 **43.** 357 **45.** – 60 **47.** 7 **49.** – 4 **51.** 4,711 **53.** – 305 **55.** 215 **57.** – 112 **59.** 3 **61.** – 6 **63.** 3 **65.** – 7 **67.** 9 **69.** – 562 **71.** 2 **73.** 0 **75.** 0 **77.** – 2 **79.** – 1 **81.** – 3

83. –1,032 **85.** – 21 **87.** – 8,348 **89.** – 20 **91. a.** 27 **b.** 3 **c.** – 27 **d.** – 3 **93. a.** answers may vary **b.** answers may vary **c.** answers may vary **95.** answers may vary **97.** 110°F,114°F **99. a.** – 15,720 ft **b.** – 12,500 ft **101. a.** – 9 ft **b.** 2 ft above flood stage **103.** 195° **105.** 4 % **107.** 3,250 m **109.** ($967) **111.** –$200 **113.** – 8, – 5, – 3, –3 **115.** –$ 99,650,000 **117.** Yes

Applications Introduction : Subtracting Integers Section 1.3 (page 37)

1. a. 9 − (−3) **b.** −20 − 11 **c.** −8 − (−4) **2. a.** 1 – (– 7) **b.** – 1 – 7 **c.** – 1 – (– 7) **d.** 1 – 7 **3.** –15 – (– 3) **4.** – 120 – (– 4) **5. a.** 22 **b.** 105 **c.** 260 **d.** 532 **e.** 1,700 **f.** 21,000 **6.** adding

Are You Ready? Section 1.3 (page 38)

1. –6, 15 **2.** 22–6 **3.** 12 **4.** – 9

Section 1.3 Study Set (page 44)

1. opposite, additive **3.** Value **5.** opposite **7.** – 3 **9.** change **11. a.** 3 **b.** – 12 **13.** 3 – (–6) = 3 + 6 = 9 **15. a.** – 8 – (– 4) **b.** – 4 – (– 8) **17.** –3, 2, 0 **19.** –2 , – 10, 6, – 4 **21.** –7 **23.** – 10 **25.** 9 **27.** 18 **29.** – 18 **31.** – 50 **33. a.** – 10 **b.** 10 **35. a.** 25 **b.** – 25 **37.** –15 **39.** 9 **41.** –2 **43.** – 10 **45.** 9 **47.** – 12 **49.** – 8 **51.** 0 **53.** 32 **55.** – 26 **57.** – 2,447 **59.** 43,900 **61.** 3 **63.** 10 **65.** 8 **67.** 5 **69.** 3 **71.** – 1 **73.** – 9 **75.** – 22 **77.** 9 **79.** – 4 **81.** 0 **83.** – 18 **85.** 8 **87.** – 25 **89. a.** – 53 **b.** – 47 **91. a.** answers may vary **b.** answers may vary **93. a.** – 15 – (– 25) = 10 **b.** – 25 – 15 = – 40 **95.** – 2,200 ft **97.** 1,030 ft **99.** – 8 **101.** – 4 yd **103.** – $140 **105.** Portland, Barrow, Greensboro, Norfolk, Atlantic City **107.** 470°F **109.** 16-point increase **111.** – 79,200, – 71,500, – 64,200 **113.** +29, +35, – 23 **115.** 428 B.C.

Are You Ready? Section 1.4 (page 48)

1. 14, 67 **2.** Different signs **3.** – 12 **4.** 66,834 **5.** 32,300 **6.** 60

Study Set Section 1.4 (page 54)

1. factor, factor, product **3.** unlike **5.** Associative **7.** negative **9.** negative **11.** unlike/different **13.** 0 **15. a.** 3 **b.** 12 **17. a.** –3(– 2) **b.** 5 (– 5) **c.** – (5 • 5) **19. a.** 8 ·5 **b.** (– 2 • 6) 9 **c.** 0 **d.** 1 **21.** –15 **23.** – 18 **25.** – 72 **27.** – 126 **29.** – 1,665 **31.** – 94,000 **33.** 56 **35.** 7 **37.** 156 **39.** 276 **41.** 1,947 **43.** 72,000,000 **45.** 90 **47.** 150 **49.** – 384 **51.** – 336 **53.** – 48 **55.** – 81 **57.** 36 **59.** 144 **61.** – 60 **63.** 0 **65.** – 441 **67.** – 20 **69.** – 18 **71.** 60 **73.** – 48 **75.** –8,400,000 **77.** – 12 **79.** 144 **81.** 54 **83.** – 120 **85.** – 216 **87.** 0 **89. a.** answers may vary **b.** answers may vary **91.** answers may vary **93.** – 5 and – 4 **95.** – 2,000 ft **97. a.** high: 2, low: – 3 **b.** high: 4, low: – 6

99. a. – 402,000 jobs **b.** – 423,000 jobs **c.** – 581,000 jobs **d.** – 528,000 jobs **101.** – 324°F **103.** – $1,200 **105.** – 18 ft **107.** – $215,718 **109.** – 72° **111.** 340, – 9,240, – 40,800

Are You Ready? Section 1.5 (page 57)

1. 45, 225 **2.** Different signs **3.** 9 **4.** 90 **5.** 218 **6.** 56

Study Set Section 1.5 (page 63)

1.

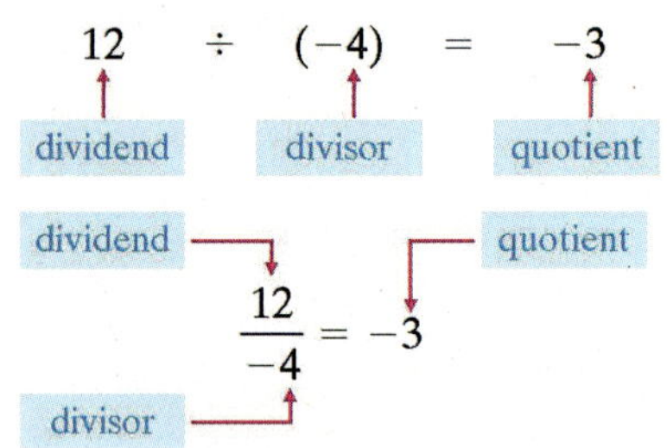

3. by, of **5. a.** – 5 (5) = – 25 **b.** 6 (–6) = – 36 **c.** 0 (–15) = 0 **7. a.** positive **b.** negative **9. a.** 0 **b.** undefined **11. a.** always true **b.** sometimes true **c.** always true **13.** – 7 **15.** – 4 **17.** – 6 **19.** – 8 **21.** – 22 **23.** – 39 **25.** – 30 **27.** – 50 **29.** 2 **31.** 5 **33.** 9 **35.** 4 **37.** 16 **39.** 21 **41.** 40 **43.** 500 **45. a.** undefined **b.** 0 **47. a.** 0 **b.** undefined **49.** 3 **51.** – 17 **53.** 0 **55.** – 5 **57.** – 5 **59.** undefined **61.** – 19 **63.** 1 **65.** – 20 **67.** – 1 **69.** 10 **71.** – 24 **73.** – 30 **75.** – 4 **77. a.** 18 **b.** 36 **c.** – 243 **d.** – 3 **79. a.** answers may vary **b.** answers may vary **81. a.** answers may vary **b.** answers may vary **83.** – 542 **85.** – 1,634 **87.** – $35 per week **89.** – 1,010 ft **91.** – 7° per min **93.** – 6 (6 games behind) **95.** – $15 **97.** – $17 **99.** – $8,000 per year

Applications Introduction: Order of Operations Section 1.6 (page 66)

1. b. 42 tickets **2. a.** Leslie Owens won **b.** John Hunt won **c.** Because of the changes in the punctuation, we read them differently.

Are You Ready? Section 1.6 (page 66)

1. Subtraction, multiplication **2.** Addition, multiplication, division **3.** 81 **4.** –125

Study Set Section 1.6 (page 75)

1. base, exponent **3.** power **5.** order **7. a.** addition and multiplication **b.** 54, 34 **c.** 34, multiplication is done before addition **9.** Multiplication **11. a.** addition, power, multiplication **b.** power, multiplication, addition **13.** multiplication, subtraction **15. a.** power **b.** addition **c.** subtraction **d.** division **e.** subtraction **17.** 12^6 **19. a.** –5 **b.** 5 **21.** 9, 54 **23.** 3, 9, 18 **25.** 3^4 **27.** $10^2 \cdot 12^3$ **29.** 49 **31.** 216 **33.** 625 **35.** –8 **37.** –27 **39.** –36 **41.** –17 **43.** 187 **45.** 30 **47.** 19 **49.** 43 **51.** 8 **53.** –34 **55.** –118 **57.** 58 **59.** 49 **61.** 1,000 **63.** –44 **65.** 201 **67.** –396

69. 50 **71.** –27 **73.** 69 **75.** –10 **77.** 10 **79.** 12 **81.** 1 **83.** –8 **85.** 31 **87.** 1 **89.** 11 **91.** –1,296 **93.** –44 **95.** 12 **97.** 8 **99.** –24 **101.** 20 **103.** 343 **105.** 360 **107.** 12 **109.** 59 **111.** 28 **113.** –24 **115.** –2 **117.** Undef. **119.** –39 **121.** –8 **123.** 10 **125.** –1 **127.** 427 **129.** **a.** 22 **b.** –13 **131.** **a.** –40 **b.** –10 **133.** **b.** 2^{10} = 1,024, 4, 2, 8, 3, 16, 4 **135.** 114 points, 3, 1, 1, 1, 1, 10, 4 **137.** 2,7,5

Are You Ready? Section 1.7 (page 79)

1. multiplication **2.** subtraction **3.** division **4.** addition **5.** subtracting 3 **6.** multiplying by 2 **7.** adding 12 **8.** dividing by 4

Study Set Section 1.7 (page 86)

1. expressions **3.** constant **5.** multiplication, division **7.** **a.** equation **b.** algebraic expression **9.** **a.** algebraic expression **b.** equation **11.** addition, multiplication, division; t **13.** **a.** $10 + x$ **b.** $3t - 2$ (answers may vary) **15.** **a.** ii **b.** iii **c.** iv **d.** i **17.** $12 - h$ **19.** 10, 20, 30, $10d$, multiply **21.** **a.** $8x$ **b.** $5t$ **c.** $\frac{10}{g}$ **d.** Prt **e.** xy **23.** $l + 15$ **25.** $50x$ **27.** $w - l$ **29.** $P + 2p$ **31.** $k^2 - 2{,}005$ **33.** $2a - 1$ **35.** $\frac{1{,}000}{n}$ **37.** $2p + 90$ **39.** $3(35 + h + 300)$ **41.** $p - 680$ **43.** $4d - 15$ **45.** $2(200 + t)$ **47.** $|a - 2|$ **49.** $10d$ **51.** $f - 2$ **53.** $6s$ **55.** $\frac{p}{15}$ **57.** $t + 2$ **59.** $\frac{h}{4}$ **61.** $450 - x$ **63.** w = the width of the rectangle (in inches); $w + 6$ = the length of the rectangle (in inches)

65. g = the number of quarts of coolant originally in the radiator; $g - 3$ = the number of quarts of coolant that are left in the radiator **67.** v = the area of Vermont (in square miles); $50v + 380$ = the area of Alaska (in square miles) **69.** s = the number of calories in a scoop of ice cream; $2s + 100$ = the number of calories in a slice of pie **71.** **a.** $b + 30$ **b.** $e - 30$ **73.** **a.** $s + 11$ **b.** $w - 11$ **75.** x = the age of the ATM; $x + 11$ = the age of the digital clock; $x - 15$ = the age of the camcorder **77.** x = the age of the Empire State Building; $x + 18$ = the age of the Woolworth Building; $x - 21$ = the age of the United Nations Building **79.** $60m$ sec **81.** $12f$ in.

83. $\frac{y}{100}$ centuries **85.** $\frac{e}{12}$ dozen **87.** the quotient of r and 3 **89.** 50 less than t **91.** the product of x, y, and z **93.** twice m, increased by 5 **95.** **a.** $7x$ hours **b.** $365x$ hours **97.** **a.** $\frac{s}{12}$ dollars **b.** $\frac{s}{52}$ dollars **99.** x = the number of votes received by Nixon; $x + 118{,}550$ = the number of votes received by Kennedy **101.** let x = the age of Apple; $x + 80$ = the age of IBM; $x - 9$ = the age of Dell **103.** 500, $500 + x$, $500 - x$ **105.** $5{,}000 - x$

Applications Introduction: Variables and Formulas Section 1.8 (page 92)

1. **a.** $P = 100 - d$ **b.** $c = m + 7$ **c.** $7d = h$ **d.** $c = \frac{y}{100}$ **e.** $t = 2{,}500 - d$ **f.** $d = \frac{y}{10}$ **g.** $c = 100h + 200$ **2.** **a.** $P = r - c$ **b.** $A = l \cdot w$ **c.** $D = 2r$ **d.** $R = s + d$ **e.** $P = l + w + l + w$

Are You Ready? Section 1.8 (page 93)

1. 103 **2.** –19 **3.** –12 **4.** 0

Study Set Section 1.8 (page 101)

1. expression **3.** substitute **5.** order **7.** **a.** $s = p - d$ **b.** $p = r - c$ **c.** $r = c + m$ **9.** 5, 25, 45 **11.** **a.** x = the length of part 1; $x - 40$ = the length of part 2; $x + 16$ = the length of part 3 (answers may vary) **b.** part 2: 20 in.; part 3: 76 in. **13.** **a.** x = the weight of a Honda Element; $2x - 340$ = the weight of an H2 Hummer; $x - 1{,}720$ = the weight of a Smart Fortwo car (answers may vary) **b.** H2 Hummer: 6,400 lb; Smart Fortwo car: 1,650 lb **15.** 27 **17.** –4 **19.** 16 **21.** –17 **23.** –51 **25.** 2 **27.** 144 **29.** 6 **31.** \$165 **33.** \$190 **35.** \$150 **37.** \$8,200 **39.** 1,650 mi **41.** 96 mi **43.** 64 ft **45.** 256 ft **47.** 239 **49.** 25 lb **51.** –6 **53.** 4 **55.** –30 **57.** 23 **59.** –3 **61.** 4 **63.** 3 **65.** 65 **67.** 44 **69.** –21 **71.** –26 **73.** 25 **75.** undefined **77.** –5 **79.** –270 **81.** 21 **83.** 6,166; 6,744 **85.** speedometer: rate; odometer: distance; clock: time; $d = rt$ **87.** **a.** 500; 26,100; 500; 23,700; 500; $x + 500$ **b.** $p = f + 500$ **89.** **a.** 414,000 mi **b.** 11,160,000 mi **c.** 33,000 ft **91.** **a.** –5, –3, –1, 15, 45 **b.** –5, $2m$ **c.** 32 min **d.** 59°C **93.** 12 min **95.** 31 therms

Applications Introduction: Perimeter and Area Section 1.9 (page 107)

1. **a.** area **b.** perimeter **c.** area **d.** area **e.** perimeter **f.** perimeter **g.** area **2.** **a.** iii **b.** vii **c.** vi **d.** v **e.** i **f.** ii **g.** iv

Are You Ready? Section 1.9 (page 107)

1. 344 **2.** 102 **3.** 225 **4.** 6 **5.** 95 **6** **a.** 3 ft **b.** 12 in. **c.** 36 in.

Study Set Section 1.9 (page 116)

1. perimeter **3.** area **5.** area **7.** 8 ft • 16 ft = 128 ft^2 **9.** **a.** $P = 4s$ **b.** $P = 2l + 2w$ **11.** **a.**

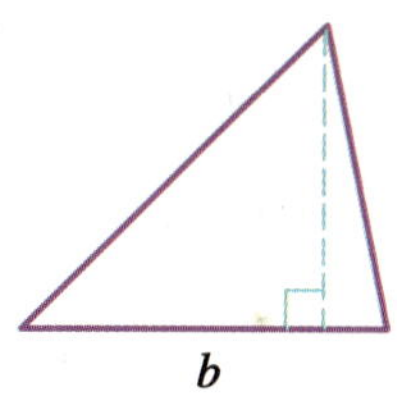

b.

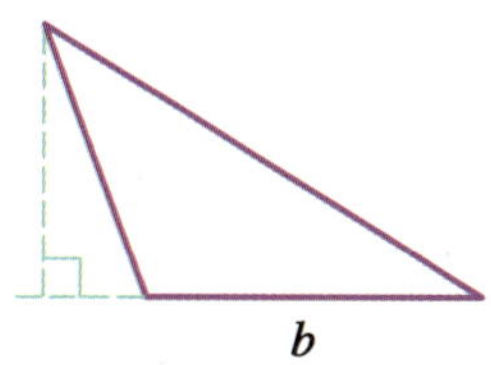

13. a rectangle and a triangle **15.** **a.** square inch **b.** 1 m^2 **17.** 32 in. **19.** 24 mi **21.** 62 in **23.** 94 in. **25.** 35 ft **27.** 36 m **29.** x = 4 in., y = 3 in.; 77 in. **31.** 16 cm^2 **33.** 625 m^2 **35.** 144 $in.^2$ **37.** 1,000,000 mm^2 **39.** 27,878,400 ft^2 **41.** 1,000,000 m^2 **43.** 135 ft^2 **45.** 11,160 ft^2 **47.** 25 $in.^2$ **49.** 30 cm^2 **51.** 5 $in.^2$

53. 162 mi^2 **55.** 102 $in.^2$ **57.** 360 ft^2 **59.** \$1,200 **61.** \$4,680 **63.** 75 m^2 **65.** 75 yd^2 **67.** length 15 in. and width 5 in.; length 16 in. and width 4 in. (answers may vary) **69.** sides of length 5 m **71.** base 5 ft and height 8 ft (answers may vary) **73.** length 4 ft and width 25 ft; length 8 ft and width 8 ft (answers may vary) **75.** 12 cm **77.** 36 m **79.** 11 mi **81.** 144 plants **83.** carpeting **85.** 6 gal **87.** 51 sheets

Are You Ready? Section 1.10 (page 121)

1. 36 **2.** 225 **3.** 9 **4.** 12

Section 1.10 Study Set (page 125)

1. square **3.** radical **5.** perfect **7.** 25, 25 **9.** **a.** 7 **b.** 2 **11. a.** 1 **b.** 0 **13.** Step 2: Evaluate all exponential expressions and any square roots. **15.** square, radical **17.** **a.** square root **b.** negative **19.** –7, 8 **21.** 5 and –5 **23.** 4 and –4 **25.** 4 **27.** 3 **29.** –12 **31.** –7 **33.** 7 **35.** 16 **37.** –16 **39.** –3 **41.** 20 **43.** –140 **45.** –48 **47.** 43 **49.** 75 **51.** –7 **53.** –1 **55.** –10 **57.** 64 **59.** –140 **61.** 31 **63.** 3 **65.** 15 **67.** 7 **69.** 1 **71.** 10 **73.** **a.** 5 ft **b.** 10 ft **75.** 84 in., 7 ft **77.** 42-inch screen **79.** 1,080 mm

Applications Introduction: The Pythagorean Theorem Section 1.11 (page 128)

1. **a.** No **b.** Yes **c.** Yes **d.** No **e.** Yes **f.** No

2. **a.** 36 + 64 = 100 **b.** 25 + 144 = 169
When you add the number of small squares in the large squares drawn on each shorter side, you get the number of small squares in the large square drawn on the longest side of the right triangle.

Are You Ready? Section 1.11 (page 129)

1. 90 **2.** 3 **3.** **a.** 5 **b.** 15 **4.** **a.** 37 **b.** 472

Study Set Section 1.11 (page 132)

1. hypotenuse, legs **3.** Pythagorean **5.** **a.** obtuse **b.** right **c.** acute **d.** straight **7.** **a.** obtuse **b.** right **c.** acute **d.** obtuse **9.** a^2, b^2, c^2 **11.** right **13.** 64, 100, 100 **15.** 10 ft **17.** 13 m **19.** 73 mi **21.** 137 cm **23.** no **25.** yes **27.** 10 yd **29.** 25 in. **31.** 17 ft **33.** **a.** 16 in., 30 in. **b.** 34 in. **35.** 53 in. **37.** 44 ft **39.** 60 in. **41.** **b.** 53 paces

Module 1 Test (page 136)

1. **a.** integers **b.** inequality **c.** absolute, value **d.** variables **e.** base, exponent **f.** expressions **g.** substitute **h.** equation **i.** formula **j.** perimeter, area **k.** isosceles **l.** radical, positive, negative **m.** right **n.** hypotenuse, legs **o.** Pythagorean **2.** **a.** > **b.** < **c.** < **3.** **a.** true **b.** true **c.** false **d.** false **e.** true **4.** Poly

5. 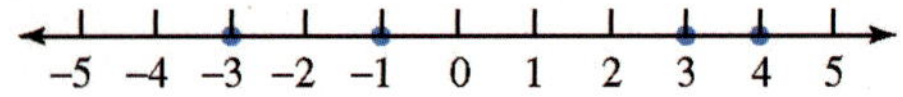

6. **a.** Always true **b.** Always true **c.** Sometimes true **d.** Never true **7.** – 21 – (– 73) **8.** **a.** 6 + (– 2) = 4 **b.** – 7 + 3 = – 4 **c.** 1 + (– 6) = – 5 **d.** – 4 + (– 5) = – 9 **9.** **a.** 0 **c.** 4 + (– 4) = 0 **10.** **a.** – 3 **b.** – 145 **c.** – 1 **d.** – 32 **e.** – 3 **11.** **a.** – 13 **b.** – 1 **c.** 191 **d.** – 15 **e.** – 150 **12.** **a.** – 70 **b.** 292 **c.** 48 **d.** 54 **e.** –26,000,000 **13.** 5 (– 4) = – 20 **14.** **a.** – 8 **b.** – 8 **c.** 9 **d.** – 34 **e.** – 80 **15.** **a.** – 12 **b.** 18 **c.** 4 **d.** – 80 **16.** **a.** answers may vary **b.** answers may vary **c.** answers may vary **d.** answers may vary **17.** – 6 and 5 **18.** **a.** commutative property of addition **b.** commutative property of multiplication **c.** adding **19.** **a.** undefined **b.** – 5 **c.** 0 **d.** 1 **20.** **a.** Associative property of addition **b.** –253 **21.** **a.** 16 **b.** –16 **22.** 1 **23.** –27 **24.** –34 **25.** 88 **26.** 6 **27.** 48 **28.** –24 **29.** 58 **30.** –72°F **31.** \$203 lost (–203) **32.** 154 ft **33.** –350 ft **34.** –15 **35.** –\$60 million **36.** **a.** –25 –25 = –50 **b.** 25 – (–10) = 35 **37.** 1,100°F **38.** –\$140 **39.** **a.** equation **b.** algebraic expression **c.** algebraic expression **d.** equation **40.** Addition, multiplication, division; t **41.** **a.** $2h - 1{,}000$ **b.** \$3,700 **42.** $56 - c$ **43.** t = the length of the trout, $t + 10$ = the length of the salmon; or s = the length of the salmon, $s - 10$ = the length of the trout **44.** **a.** $r - 2$ **b.** $3xy$ **c.** $x + 100$ **d.** $\left|\frac{x}{-9}\right|$ **45.** $10d$ **46.** **a.** –12 **b.** 26 **c.** 1 **d.** 23 **47.** The distance traveled is 165 mi. **48.** The profit is \$37,000. **49.** It would be 56 ft short of hitting the ground. **50.** **a.** Let x = weight of the Element, $2x - 340$ = weight of the Hummer **b.** 6,400 lb **51.** The mean meter reading is 1. **52.** **a.** Area; ft^2 **b.** Perimeter; ft **53.** 148 in. **54.** 3,600 ft^2 **55.** 120 $in.^2$ **56.** \$1,120 **57.** 360 cm^2 **58.** 32 yds, 34 yds **59.** **a.** length: 9 ft, width: 10 ft; length: 30 ft, width: 3 ft (answers may vary) **b.** No **60.** 238 cm^2 **61.** 56 m^2 **62.** **a.** 11 **b.** –12 **63.** 9 sec **64.** 26 m **65.** 29 in. **66.** 13 ft **67.** **b.** 157 ft **68.** yes